AF347526

Industrial Wireless Sensor Networks

Industrial Wireless Sensor Networks: Protocols and Applications

Editors

Seong-eun Yoo
Taehong Kim

MDPI • Basel • Beijing • Wuhan • Barcelona • Belgrade • Manchester • Tokyo • Cluj • Tianjin

Editors

Seong-eun Yoo
Daegu University
Korea

Taehong Kim
Chungbuk National University
Korea

Editorial Office
MDPI
St. Alban-Anlage 66
4052 Basel, Switzerland

This is a reprint of articles from the Special Issue published online in the open access journal *Sensors* (ISSN 1424-8220) (available at: https://www.mdpi.com/journal/sensors/special_issues/IWSNPA).

For citation purposes, cite each article independently as indicated on the article page online and as indicated below:

LastName, A.A.; LastName, B.B.; LastName, C.C. Article Title. *Journal Name* **Year**, *Article Number*, Page Range.

ISBN 978-3-03943-605-7 (Hbk)
ISBN 978-3-03943-606-4 (PDF)

Contents

About the Editors

Seong-eun Yoo received a B.S. degree in electronics and computer engineering from Hanyang University, Seoul, South Korea, in 2003, and M.S. and Ph.D. degrees in information and communications engineering from Korea Advanced Institute of Science and Technology (KAIST), Daejeon, South Korea, in 2005 and 2010, respectively. Since September 2010, he has been a faculty member with the School of Computer and Communication Engineering, Daegu University, Gyeongsan, South Korea. His research interests include real-time communication in wireless sensor networks and Internet of Things, as well as real-time embedded systems.

Taehong Kim received his B.S. degree in computer science from Ajou University, South Korea, in 2005, and his M.S. degree in information and communication engineering from Korea Advanced Institute of Science and Technology (KAIST) in 2007. He received his Ph.D. degree in computer science from KAIST in 2012. He worked as a research staff member at Samsung Advanced Institute of Technology (SAIT) and Samsung DMC R&D Center from 2012 to 2014. He also worked as a senior researcher at the Electronics and Telecommunications Research Institute (ETRI), South Korea, from 2014 to 2016. Since 2016, he has been an associate professor with the School of Information and Communication Engineering, Chungbuk National University, South Korea. He has been an associate editor of IEEE access since 2020. His research interests include edge computing, SDN/NFV, the Internet of Things, and wireless sensor networks.

Editorial

Industrial Wireless Sensor Networks: Protocols and Applications

Seong-eun Yoo [1] and Taehong Kim [2,*]

[1] School of Computer and Communication Engineering, Daegu University, Gyeongsan 38453, Korea; seyoo@daegu.ac.kr

[2] School of Information and Communication Engineering, Chungbuk National University, Cheongju 28644, Korea

* Correspondence: taehongkim@cbnu.ac.kr; Tel.: +82-43-261-2481

Received: 12 October 2020; Accepted: 13 October 2020; Published: 14 October 2020

Abstract: Wireless sensor networks are penetrating our daily lives, and they are starting to be deployed even in an industrial environment. The research on such industrial wireless sensor networks (IWSNs) considers more stringent requirements of robustness, reliability, and timeliness in each network layer. This Special Issue presents the recent research result on industrial wireless sensor networks. Each paper in the special issue has unique contributions in the advancements of industrial wireless sensor network research and we expect each paper to promote the relevant research and the deployment of IWSNs.

Keywords: protocols for Industrial Wireless Sensor Networks (IWSN); IWSN standards; IEEE 802.15.4; WirelessHART; ISA100.11a; real-time communications in IWSN; Wireless Networked Control Systems (WNCSs); IWSN testbeds and applications

---

1. Introduction

With the help of technological advances and standardization activities, we are witnessing the growth of wireless sensor network applications in many areas, such as home appliances, agriculture, transportation, and manufacturing. Recently, the adoption of wireless sensor networks (WSNs) in industrial areas has increased due to matured technologies such as transceivers and protocols. Industrial wireless sensor networks (IWSNs) require reliability, robustness, and timeliness in the information exchange between devices, and they need more sophisticated protocols that many researchers and developers are working on.

This Special Issue focuses on the latest research, application, and adoption of wireless sensor networks in industrial fields from the perspective of protocols and applications, requiring high reliability and real-time packet delivery. In such a context, this Special Issue invited contributions in the following topics (though without being limited to them):

- Protocols for industrial wireless sensor networks
- Wireless network control systems
- Real-time communications in IWSNs
- IWSN testbeds and applications such as smart factories
- Edge and fog computing for IWSN
- Any subjects relevant to IWSN.

There were number of submissions and two to four reviewers evaluated each submitted article and finally six outstanding papers were accepted and published in this Special Issue. We present the brief summary of each accepted paper in the following section.

2. Brief Review of the Published Articles

The first paper [1] on the network layer presents an adaptive real-time routing protocol for an (m, k)-firm real-time model. The model defines the real-time requirement that at least m out of any k consecutive messages from a real-time stream must meet their deadlines to ensure an adequate Quality of Service (QoS). The paper proposes a new adaptive path selection and traffic shaping algorithm. The evaluation in OPNET shows that the proposed scheme can reduce the dynamic failure regardless of the background traffic while reducing the energy consumption.

The next two papers are relevant to Media Access Control (MAC) sub-layer. To provide a remedy for the joint problem of the low-access efficiency and the interference diffusion in high-density deployment of IWSNs, the second paper [2] on MAC (Media Access Control) proposes a spatial group-based multi-user Full Duplex OFDMA (GFDO) protocol. While the theoretical analysis derives the average number of nodes in an access channel, system saturation throughput, and area throughput, the simulation in an NS2 simulator supports the theoretical analysis and shows the efficiency of the proposed protocol.

Another paper [3] related to the MAC sub-layer issue proposes a novel distributed scheduling scheme for an ad-hoc network. The proposed scheme consists of two steps of reallocation procedures: intra-node and inter-node. The intra-node slot scheduling reallocates the packets in priority-based multiple queues using a self-fairness index to increase the throughput and delay, while the inter-node scheduling reallocates slots between neighboring nodes to increase fairness. The simulation study with a Java-based network simulator shows that the proposed scheme can adjust the packet delivery performance according to a predefined weight factor and outperform the conventional algorithms in throughput and delay.

Time synchronization is another important research area in IWSNs. The paper on the time synchronization issue proposes an energy-efficient reference node selection (EERS) algorithm [4]. EERS minimizes the number of connected reference nodes and keeps a minimal number of hops. The evaluation in an experimental network with 25 real hardware platforms shows EERS needs fewer messages (i.e., less energy consumption) than the existing techniques such as R-Sync [5], FADS [6], and LPSS [7]. In addition, the simulation study with MATLAB for a larger scale network evaluation shows that the proposed protocol provides energy efficiency without losing the accuracy of the time synchronization.

There are two papers related to the popular IWSN standards: IEEE 802.15.4 and ISA 100.11a. The first paper [8] considers the event monitoring in hazards and disaster detection in a large area with the IEEE 802.15.4/ZigBee standard. This paper proposes a dynamic reconfiguration mechanism of cluster-tree IWSNs. The proposed algorithm can dynamically reconfigure large-scale IEEE 802.15.4 cluster-tree IWSNs to assign communication resources to the overloaded branches of the tree based on the accumulated network load generated by each of the sensor nodes. The simulation assessment with CT-Sim [9] shows that the proposed scheme can guarantee the required quality of service level for the dynamic reconfiguration of cluster-tree networks.

The last paper [10] presents a wireless network control system (WNCS) composed of ISA 100.11a-based field devices (i.e., sensor nodes), a network manager, a controller, and a wired actuator. The system controls the liquid level in the tank of the coupled tank system. In order to assess the influence of the sensor link failure on the control loop, the controller calculates the link stability and chooses an alternative link in case of instability in the current link. Preliminary experimental tests of WNCS performance shows that the link stability factor allows a prediction of the change in the control system error. Finally, the tests of the control system based on link stability show that the link stability metric is able to identify possible instabilities and prevent the failure of the system's control loop. Even with disturbances in the network links, the control system error remains below the threshold.

Funding: This work was supported by the Daegu University Research Grant (No. 20170034).

Acknowledgments: The authors would like to thank all the authors for their submissions to this special collection. The authors are also grateful to the reviewers for contributing long hours in reviewing papers and submitting their assessments in a professional and timely manner.

Conflicts of Interest: The authors declare no conflict of interest.

References

1. Kim, B.-S.; Kim, S.; Kim, K.H.; Sung, T.-E.; Shah, B.; Kim, K.-I. Adaptive Real-Time Routing Protocol for (m,k)-Firm in Industrial Wireless Multimedia Sensor Networks. *Sensors* **2020**, *20*, 1633. [CrossRef] [PubMed]
2. Peng, M.; Li, B.; Yan, Z.; Yang, M. A Spatial Group-Based Multi-User Full-Duplex OFDMA MAC Protocol for the Next-Generation WLAN. *Sensors* **2020**, *20*, 3826. [CrossRef] [PubMed]
3. Lee, W.; Kim, T.; Kim, T. Distributed Node Scheduling with Adjustable Weight Factor for Ad-hoc Networks. *Sensors* **2020**, *20*, 5093. [CrossRef] [PubMed]
4. ElSharief, M.; El-Gawad, M.A.A.; Ko, H.; Pack, S. EERS: Energy-Efficient Reference Node Selection Algorithm for Synchronization in Industrial Wireless Sensor Networks. *Sensors* **2020**, *20*, 4095. [CrossRef] [PubMed]
5. Qiu, T.; Zhang, Y.; Qiao, D.; Zhang, X.; Wymore, M.L.; Sangaiah, A.K. A Robust Time Synchronization Scheme for Industrial Internet of Things. *IEEE Trans. Ind. Inform.* **2017**, *14*, 3570–3580. [CrossRef]
6. ElSharief, M.; El-Gawad, M.A.A.; Kim, H. FADS: Fast Scheduling and Accurate Drift Compensation for Time Synchronization of Wireless Sensor Networks. *IEEE Access* **2018**, *6*, 65507–65520. [CrossRef]
7. ElSharief, M.; Abd El-Gawad, M.A.; Kim, H. Low-Power Scheduling for Time Synchronization Protocols in A Wireless Sensor Networks. *IEEE Sens. Lett.* **2019**, *3*, 1–4. [CrossRef]
8. Lino, M.; Leão, E.; Soares, A.; Montez, C.; Vasques, F.; Moraes, R. Dynamic Reconfiguration of Cluster-Tree Wireless Sensor Networks to Handle Communication Overloads in Disaster-Related Situations. *Sensors* **2020**, *20*, 4707. [CrossRef]
9. Leão, E.; Moraes, R.; Montez, C.; Portugal, P.; Vasques, F. CT-SIM: A simulation model for wide-scale cluster-tree networks based on the IEEE 802.15.4 and ZigBee standards. *Int. J. Distrib. Sens. Netw.* **2017**, *13*, 1–17.
10. Florencio, H.; Dória Neto, A.; Martins, D. ISA 100.11a Networked Control System Based on Link Stability. *Sensors* **2020**, *20*, 5417. [CrossRef] [PubMed]

Publisher's Note: MDPI stays neutral with regard to jurisdictional claims in published maps and institutional affiliations.

Article

Adaptive Real-Time Routing Protocol for (*m*,*k*)-Firm in Industrial Wireless Multimedia Sensor Networks [†]

Beom-Su Kim [1], Sangdae Kim [1], Kyong Hoon Kim [2], Tae-Eung Sung [3], Babar Shah [4] and Ki-Il Kim [1,*]

[1]   Department of Computer Science and Engineering, Chungnam National University, Daejeon 34134, Korea; bumsou10@cnu.ac.kr (B.-S.K.); sdkim.cse@gmail.com (S.K.)

[2]   School of Computer Science and Engineering, Kyungpook National University, Daegu 41566, Korea; kyong.kim@knu.ac.kr

[3]   Department of Computer and Telecommunications Engineering, Yonsei University, Wonju 26493, Korea; tesung@yonsei.ac.kr

[4]   College of Technological Innovation, Zayed University, P.O. Box 144534, Abu Dhabi, UAE; Babar.Shah@zu.ac.ae

*   Correspondence: kikim@cnu.ac.kr

[†]   This paper is an extended version of our paper published in 7th International Symposium on Sensor Science (I3S 2019), Naples, Italy, 9–11 May 2019.

Received: 16 February 2020; Accepted: 12 March 2020; Published: 14 March 2020

Abstract: Many applications are able to obtain enriched information by employing a wireless multimedia sensor network (WMSN) in industrial environments, which consists of nodes that are capable of processing multimedia data. However, as many aspects of WMSNs still need to be refined, this remains a potential research area. An efficient application needs the ability to capture and store the latest information about an object or event, which requires real-time multimedia data to be delivered to the sink timely. Motivated to achieve this goal, we developed a new adaptive QoS routing protocol based on the (*m*,*k*)-firm model. The proposed model processes captured information by employing a multimedia stream in the (*m*,*k*)-firm format. In addition, the model includes a new adaptive real-time protocol and traffic handling scheme to transmit event information by selecting the next hop according to the flow status as well as the requirement of the (*m*,*k*)-firm model. Different from the previous approach, two level adjustment in routing protocol and traffic management are able to increase the number of successful packets within the deadline as well as path setup schemes along the previous route is able to reduce the packet loss until a new path is established. Our simulation results demonstrate that the proposed schemes are able to improve the stream dynamic success ratio and network lifetime compared to previous work by meeting the requirement of the (*m*,*k*)-firm model regardless of the amount of traffic.

Keywords: (*m*,*k*)-firm model; adaptive real-time routing; traffic management; industrial wireless multimedia sensor network

---

1. Introduction

Even though many typical Wireless Sensor Networks (WSNs) have been employed to detect the events and report interested values around itself, they are typically based on scalar values and therefore have limited abilities to obtain various types of information. This implies that limited information, such as the geographic position of the target, can be recorded. As an alternative, wireless multimedia sensor networks (WMSN) [1] can capture multimedia information about the object or event by using a camera and microphone. However, the more complicated nature of multimedia data (compared to scalar values) has prompted considerable research in an attempt to effectively capture the information as well as to deliver this information in real time. One of the good potential examples of WMSN is industrial sensor networks [2] to monitor and control systems. In industrial sensor networks, it is essential to employ QoS routing protocol in WMSN to accomplish the mission.

To address the problems associated with real-time delivery in WMSNs, several quality of service (QoS) routing protocols [3–5] have been proposed and discussed in the literature. However, most existing schemes still need to be improved by using a specific traffic model and applications. Based on general application model, many of them lack of applicability and lead to failure of QoS guarantee. To defeat this problem, the (m,k)-firm traffic model was recently presented for WMSNs. As a firm real-time model called (m,k)-firm is proposed to measure real-time application performance. The concept of (m,k)-firm is defined that a real-time message stream is considered to have an (m,k)-firm guarantee requirement that at least m out any k consecutive messages from the stream must meet their deadlines to ensure adequate QoS. However, the (m,k)-firm model has not been studied with respect to finding specific applications in which this model could be utilized. Thus, they have the same problem as general QoS routing protocol.

These issues motivated us to develop the adaptive real-time routing protocol we present in this paper. The model is based on the (m,k)-firm model [6,7], which serves the purpose of adding real-time capabilities. Unlike previous work, which focused on real-time delivery only, our previous approach is to introduce a multimedia application modeled by a (m,k)-firm stream. This new stream is maintained to ensure quality by using a new QoS routing protocol for the (m,k)-firm stream to reduce the overhead for path establishment and maintenance. To achieve this goal, based on the current status of application of the (m,k)-firm stream, a new QoS routing protocol was designed to select the path adaptively. When this routing protocol is unable to meet the requirement, an adaptive traffic-handling algorithm is employed. The operations of these two functions, which are dependent on the current status of the (m,k)-firm streams, ensure that high adaptability is achieved. Finally, extensive simulation results show that the proposed scheme meets the requirements of the (m,k)-firm stream for application and also provides robustness in terms of network traffic.

The main contributions of this paper are summarized below.

- To recognize the shortcomings or limitations of QoS routing protocol in WMSN, we provide a comprehensive survey that investigates state-of-the-art research work and further research directions for applying application specific requirement to the QoS routing protocol.
- To overcome the problem of general routing protocol considering deadline, we take (m,k)-firm based routing protocol while considering application requirements which is modeled by (m,k)-firm stream together.
- Unlike most of existing approaches which are based on (m,k)-firm stream, we propose a new adaptive path selection and traffic shaping algorithm.
- The proposed scheme is compared to existing (m,k)-firm based QoS routing protocols. Through various simulations, we prove that the proposed approach can reduce the dynamic failure ratio without regard to the background traffic as well as extend network lifetime.

The remainder of this paper is organized as follows. In Section 2, we discuss existing research on WMSNs and real-time routing protocols. In Section 3, we explain the real-time model, new QoS routing protocol, and traffic handling scheme. The results of the performance evaluation are provided in Section 4. Finally, the proposed scheme is summarized and further work is mentioned in Section 5.

2. Related Work

In this section, we present the related work for real-time routing protocols in WMSN. In addition to WMSN introduction, we present the outstanding potential applications including object tracking, in WMSN. Furthermore, previous real-time routing protocols are analyzed.

2.1. Wireless Multimedia Sensor Networks and Its Applications

In WMSN, each sensor node is equipped with visual and audio information collection modules such as microphones and video cameras. These nodes can retrieve multimedia content from the networks at variable rates and deliver the captured information through multi-hop communication to the sink. As compared to typical WSN, there are many research challenges that must be addressed such as high packet loss rate, shared channel, limited bandwidth, high variable delays, and lack of fixed infrastructure in WMSN. To overcome the mentioned deficiency, lots of research have been conducted in the aspects of algorithms, protocols, and hardware for the development of WMSN. Moreover, WMSN is expected to be employed increasingly for the potential applications such as surveillance and monitoring applications.

Among them, object tracking, which has been widely used and deployed in many applications, consists of object detection, object classification, and object reporting schemes. Considering these functions, wireless sensor networks (WSNs) become the potential architecture on which to implement object tracking. This is because WSNs are capable of detection and delivery at the same time in a cost efficient way. In addition, there are many applications on the WMSN such as road traffic monitoring, health-care applications, and area monitoring, etc. Road traffic monitoring [8,9] is defined as monitoring the behavior of vehicles on the road. Road traffic forecasting is important to the safety of the driver, traffic management and also fuel saving. As a promising application for real-time application, power plant [10] and port docks [11] can be monitored through deploying wireless multimedia sensor networks as well as real-time streaming to deal with emergency. Through the health-care applications [12,13], doctors and caregivers could monitor the patients' condition and activities, remotely. These two tracking and monitoring application can be applied into industrial environment by preventing unexpected events and accidents in an efficient way. As an example, Dobslaw et al. [14] proposed a new SchedEx-GA, integrated cross-layer framework, to admit a network configuration. Moreover, Sun et al. [15] proposed end-to-end data delivery reliability (E2E-DDR) to estimate and optimize the reliability in industrial sensor networks. They introduce a new mapping function with the packet reception ratio, background noise, and received signal strength by modeling background noise and path loss model. On the other hand, Zhang et al. [16] proposed an energy-efficient QoS routing algorithm along reliable path by classifying the industrial sensing data into three data types and setting their priority. This classification is used to establish forwarding node set and strategies to guarantee QoS in the aspects of reliability and real-time.

There are some research work to enhance the performance of WMSN. At first, Sarisaray-Bolukand and Akkaya [17] presented how to utilize background subtraction and compression techniques to reduce data. They analyzed the previous work in the aspects of computation and communication energy, time, and quality and provide quantitative experimental results for packet loss on the Android platform. In addition, Bavarva et al. [18] proposed how to improve system performance through multiple input multiple output along with compressive sensing. With their help, reduced energy consumption and acceptable image quality are achieved.

2.2. Real-Time Routing Protocols

The real-time transmission is essential for various applications such as multimedia, disaster response, and various studies have been proposed according to the demand. To achieve real-time transmission, the packets should be delivered to their destination within the required time of the applications. In this section, we describe previous work related to real-time routing in WSNs.

SPEED [19] has proposed a spatiotemporal approach for the real-time service. SPEED could satisfy the real-time requirement by maintaining a desired delivery speed across the networks. The desired delivery speed depends on not only the end-to-end distance but also the desired time. The intermediate node relays the packet through the node which satisfy the desired speed. As an extension of SPEED, SPEED-Realtime Routing (RR) [20] has been proposed to comprise between real-time and energy cost. To decide the next hop, three metrics, that is, residual energy of two-hop neighbor nodes, routing void and rate void rate are taken. Furthermore, adaptive control for data rate and cache queue length is employed to distinguish the congestion control. ORRP [21] presented an opportunistic real-time routing scheme. ORRP broadcasts the packet to the neighbor nodes which satisfy the real-time requirement. All neighbors received packet could obtain the priority according to a tolerable time period to be able to satisfy the requirement. As the ORRP could offer the transmission opportunity more neighbors, it could achieve reliable real-time packet transmission. Yang et al. [22] addressed that the remaining time gain could exploit to satisfy the real-time requirement. In the transmission process, the intermediate node accumulates the remaining time to gain through selecting nodes that have a higher speed than the desired speed. The accumulated time would be exploited to pass the area which no neighbors meet the real-time requirement. Unlike the previous work based on single path approach, Hassen et al. [23] presented a comprehensive multipath routing approach for QoS in WMSN. Their main strategy for survey was design issues in the aspects of multimedia data. Park et al. [24] presented a disjointed multipath routing scheme for real-time data transmission in WMSN through hybrid approach with Bluetooth and Zigbee to overcome the bandwidth limitation. Moreover, Deepaa and Sugunab [25] proposed Optimized QoS-based Clustering with Multipath Routing Protocol (OQoS-CMRP) to determine cluster heads while considering sink coverage area and energy. In addition, Kim et al. [26] presented the opportunistic multipath routing to improve the performance in WSN which can be applied to multimedia data delivery.

However, these approaches do not include real-time delivery of captured multimedia data and would need to be extended. In this regard, Kim and Sung [27] proposed cross layered approach for (m,k)-firm stream in WSN. In each layer, transmission power, priority of packet and QoS path is adjusted and found. Moreover, Li and Kim [28] presented (m,k)-firm based routing protocol which is based on a set of fault recovery mechanisms to overcome inherent resource constraint of sensor nodes and instability of wireless communication. A local status indicator to monitor and evaluate their local conditions is introduced on intermediate node. Kim et al. analyzed the challenges presented by real-time communication [29] as part of special issue [30]. Specifically, they introduced a firm real-time model together with hard and soft models. As representative examples, Kim and Sung [31] presented a distributed, measurement-based scheme for the (m,k)-firm stream. In addition, they extended this approach to address the scalability problem and lack of architecture for the (m,k)-firm stream by using flow aggregation based on a compositional hierarchical model, a velocity-based routing protocol, a hybrid medium access control protocol, and a congestion control scheme [32]. Azim et al. [33] presented a new real-time routing protocol for a (m,k)-firm model by introducing multiple attributes to evaluate the path. In addition to this, Park et al. [34] proposed a new tree-based broadcast protocol to meet the requirements of the (m,k)-firm stream in WSNs. They focused on energy efficiency by selecting as few as nodes on a tree as possible and a fast recovery scheme based on (m,k)-firm real-time conditions and local states monitoring. However, these existing schemes focus on the real-time routing based on networks environments without the considering the application requirements. This implies that a path is selected by depending on Distance Based Priority (DBP) value of (m,k)-firm stream without the quality of application.

The analysis we provide above indicates that most previous schemes would need to be extended to enable real-time delivery. Furthermore, a multimedia traffic model for streaming application would need to be introduced. Additionally, the severely constrained capabilities of sensor nodes in terms of their energy, memory, and processing power demand an adaptive approach to meet the requirements of varying network parameters.

3. Adaptive Real-Time Routing Protocol

In this section, we describe our real-time model for application and the adaptive QoS routing protocol to enable real-time processing.

3.1. Real-Time Model

We aim to model the application of multimedia data traffic and a real-time routing protocol to deliver traffic timely. We assume the event happens continuously and is recognized by a node. We focus on traffic modeling and real-time delivery. The network architecture, including the nodes, is illustrated in Figure 1, which also shows that a sink node exists in the network. Traffic model and requirements are as follows.

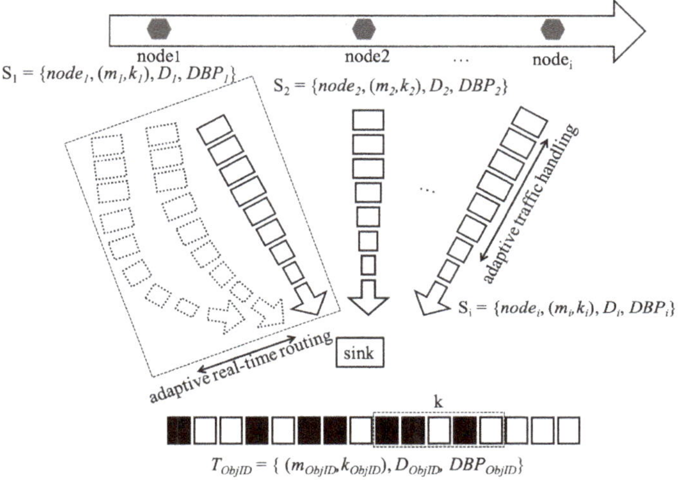

Figure 1. Overview of the proposed real-time model.

A node is capable of processing consecutive images and transmitting them to the sink node with the identity unique *ObjID*. The application stream, T_{ObjID}, is modeled by the (m,k)-firm model, that is, $T_{ObjID} = \{ m_{ObjID}, k_{ObjID}, D_{ObjID}, DBP_{ObjID}\}$ where m_{ObjID} and k_{ObjID} represent the requirements of a multimedia application. In addition, if the packet is transmitted within the deadline, D_{ObjID}, the corresponding packet is marked as Hit; otherwise, it is marked as Miss. Here, DBP_{ObjID} is prioritized value based on the distances of the stream requirement to recognize the current (m,k)-firm status. For example, with a (3,5)-firm real-time stream requirement, the traces of the reception status of the last 5 packets can be represented as HHMMM where H and M represent the Hit and Miss of status of an individual packet, respectively. Then, the DBP value becomes -1 as the required number of Hit is 3 where the Hit status remains only 2 as explained in our previous paper [35].

In addition to application stream, a real-time session flow between node i and the sink, denoted by S_i, is defined as $S_i = \{node_i, m_i, k_i, D_i, DBP_i\}$ like T_{ObjID}. The different element between T_{ObjID} and S_i is node identifier, $node_i$. Even though two model consists with the same parameters, the value for them is different according to requirements. T_{ObjID} represents the application requirement while S_i does requirement for routing protocol.

Before an event end at node i, S_i is maintained and packets are transmitted to the sink by filling T_{ObjID} with them through multiple paths as illustrated in dotted rectangle in Figure 1. According to current status of D_{ObjID} and DBP_i, one of multiple paths are selected to transmit the packets. Each session is used to partially build T_{ObjID}. Each session has the same (m_i, k_i) requirements in the T_{ObjID}. The same values for D_i are assumed for T_{ObjID} and S_i. As illustrated in Figure 1, T_{ObjID} consists of each consecutive sessions, that is $T_{ObjID} = \{ S_1 \cup S_2 \cup \ldots \cup S_n \}$. Thus, DBP_{ObjID} is affected by DBP_i as well as (m_{ObjID}, k_{ObjID}) value.

To meet the requirements of the two (m,k)-firm models, the values are assigned to comply with the following conditions. These two conditions guarantee that the requirement of S_i cannot exceed that of the T_{ObjID}. This is because, occasionally, the network cannot ensure the time requirement because of severe constraints.

$$\frac{m_i}{k_i} > \frac{m_{ObjID}}{k_{ObjID}} \tag{1}$$

$$k_i \geq k_{ObjID} \tag{2}$$

Upon detecting any event, a node i builds a multimedia stream by using consecutive packets with the (m,k)-firm model and delivers them to the sink. The sink checks whether a packet is lost or whether the packet has arrived after the deadline. These measurements are used to compute the values of DBP_i and DBP_{ObjID}, which are then transmitted to the node i.

3.2. Adaptive QoS Routing Protocol

An adaptive approach to QoS routing is taken according to DBP_i in S_i as well as DBP_{ObjID} in T_{ObjID} to meet the requirements of the (m,k)-firm in terms of both T_{ObjID} and S_i. The QoS routing algorithm is presented in Algorithm 1.

Algorithm 1 determines a suitable path for the next packets among multiple paths. To find the paths, the node sends a probing packet to identify the available paths and waits for the end-to-end delay from the sink. The probing packet is broadcasted in the same as the Route Request message (RREQ) and Route Response (RREP) in ad hoc on-demand multipath distance vector (AOMDV) [36]. The difference between the proposed and AOMDV comes from the deadline. If an intermediate node receives the packet with greater elapsed time than deadline, D_i, it does not broadcast the packet any more. In addition, the sink node replies only when the end-to-end delay is bounded within D_i. Multiple paths meeting the delay constraint are arranged in ascending order and are denoted by $P_i^{1 \cdots n}$. Based on the delay, the shortest path is denoted by P_i^1. Each node has a different maximum number of available paths, n. Initially, P_i^n is selected according to the longest-path-first strategy. After the initial stage, a path is selected by DBP_i in S_i as well as DBP_{ObjID} in T_{ObjID}. The path selection is accomplished whenever a reply with the value of DBP_{ObjID} is received.

If the value of DBP_{ObjID} is positive (lines 5–13): First, if DBP_{ObjID} is larger than 0, the application requirement is regarded to be met. So, it is important to examine the value of DBP_i. If DBP_i is larger than 0, application and session requirement are met. Thus, this path is suitable and complies with the two requirements. Otherwise, it is necessary to consider a new path with shorter delay among multiple paths, denoted by P_i^{n-1} as in line 9. However, the current path needs to be maintained if no other available path remains, when $n \leftarrow 1$. As a result, if application requirement is met, a path is re-selected according to DBP_i in S_i.

If the value of DBP_{ObjID} is negative (lines 14–29): When the value of DBP_{ObjID} is negative at node i, more complicated approaches than previous case are taken. This implies that additional options are taken into consideration to choose the path. That is, a new path is selected in aggressive way according to distance to positive status. To determine the distance to reach a positive status, we compute the absolute value with two DBP values. The larger value is obtained, the shorter path is taken. Similar to the previous case in which DBP_i was positive, an incremental approach is taken for path selection simply. On the other hand, if the value of DBP_i is also negative, we compare the absolute value with the threshold. If this value is larger than the threshold, the shortest path, P_i^1, is selected because current path is not suitable at all. This is to turn negative DBP value to positive one by ensuring the shortest delivery time between node i and the sink. Otherwise, another path is selected by choosing the path number, $n - t$, rather than $n - 1$ as in the previous case. By applying $n - t$ rather than $n - 1$, it is possible to turn negative DBP to positive one in earlier than incremental approach.

We assume that the a new event on next node, j, happen after event time on node i is over. When the new node j detects the events, Algorithm 1 is initiated on a new node. However, the path selection algorithm takes some time to execute because multiple paths should be established through

Path_Search_Probing(i,sink). To reduce the waiting time for a new path, node j temporarily sends packets directly to the previous node i until the new path is established. Node i transmits the corresponding packets along the path P_i^1. Upon establishing a path to node j, packets are delivered along P_j^n.

Algorithm 1 Adaptive QoS routing protocol for packet transmission from node i to the sink.

1: $n \leftarrow$ Number of available real-time paths
2: $P_i^{1\ldots n} \leftarrow$ Path between node i and the sink where its end-to-end delay is bounded within D_i ▷ P_i^1

    has the shortest delay whereas P_i^n has the longest delay, $P_i^1 \leq P_i^2 \leq \ldots \leq P_i^n$

3: $n \leftarrow$ Path_Search_Probing(i,sink) ▷ Obtain number of possible multiple paths from node i and sink
4: Transmit packets along P_i^n ▷ Packets are transmitted along the longest end-to-end delay in $P_i^{1\ldots n}$

5: **for** every DBP_{ObjID} **do**   ▷ After obtaining DBP_{ObjID} value of T_{ObjID}
6:     **if** $DBP_{ObjID} \geq 0$ **then**   ▷ If application stream meets the requirement
7:         **if** $DBP_i < 0$ **then**   ▷ If the session stream is not meeting the requirment
8:             **if** $n - 1 > 0$ **then**
9:                 Transmit packets along P_i^{n-1} ▷ Change the path along alternative with the next

    longest delay if possible
10:             **else**
11:                 Maintain P_i^n ▷ Current path is maintain if no alternative is available
12:             **end if**
13:         **end if**
14:         **else**   ▷ If application stream doesn't meet the requirement, more complex path selection

    algorithm is performed
15:             $t \leftarrow |DBP_{ObjID} + DBP_i|$ ▷ Compute how many distance from positive status with two

    DBP values
16:             **if** $DBP_i \geq 0$ **then**   ▷ if current session meets the requirement
17:                 **if** $n - 1 > 0$ **then**   ▷ if an available path exists
18:                     Transmit packets along P_i^{n-1} ▷ Change alternative path among multiple paths.

    Adaptive approach in the aspects of session
19:                 **else**
20:                     Transmit packets along P_i^1 ▷ Change path with the shortest end-to-end delay in case

    of two requirements missing
21:                 **end if**
22:             **else**▷ if current session and application don't meet the requirement at the same time. This is

    the worst case
23:                 **if** $t <$ Threshold **then**   ▷ Distance is not bounded within the threshold
24:                     Transmit packets along P_i^1 ▷ Change the path with the shortest end-to-end delay
25:                 **else**
26:                     Transmit packets along P_i^{n-t} ▷ Change the $(n-t)^{th}$ path according to distance to

    meeting status
27:                 **end if**
28:             **end if**
29:     **end if**
30: **end for**

3.3. Adaptive Traffic Handling

An adaptive routing algorithm ensures the selection of a path with shorter end-to-end delay than the previous path according to both DBP_i in S_i and DBP_{ObjID} in T_{ObjID}. However, if no other path is available, that is, P_i^1 is set to the current path, the adaptive traffic handling algorithm executes. The algorithm aims to reduce end-to-end delay by either redefining the (m_i, k_i)-firm model dynamically or by reducing the packet size adaptively. Without regard to adaptive handling, the application requirement remains the same. Adaptive traffic handing consists of two major functions, multiple sets for (m,k)-firm stream and dynamic payload adjustment.

First, S_i is redefined as $\{node_i, m_i, k_i, D_i, DBP_i\}$ where $(m_i, k_i) \in \{m_i^1, k_i^1\}, \ldots (m_i^s, k_i^s\}$ as shown in Algorithm 2. By redefining the (m,k)-firm stream as sets, a (m,k)-firm stream is composed with loose (m,k)-firm which conforming the initial requirement. Because this procedure is performed whenever no more path is available in adaptive routing algorithm, it aims to take loose (m,k)-firm requirement. In S_i, the lower index has, the looser (m,k)-firm requirement set. In order to maintain consistency for requirement, each element in S_i should keep the condition of Equations (1) and (2). One of the elements in the set (m_i, k_i) is selected by comparing the DBP value at consecutive moments in time, i.e., at t and $t + 1$. The selection procedure is dependent on the DBP value.

Algorithm 2 Adaptive Traffic Handling at node i.

1:   $DBPT_i^t \leftarrow$ DBP value of node i at time t
2:   $e \leftarrow$ Current index in $\{m_i^1, k_i^1), \ldots (m_i^s, k_i^s\}$ ▷ The larger index means the looser (m,k)-firm requirement
3:   $k \leftarrow$ Parameter to reduce packet size

4:   **if** $DBPT_i^t < 0$ **then**
5:     **if** $DBPT_i^{t+1} < 0$ **then**   ▷ if two consecutive DBP values are negative
6:       **if** $e\ != s$ **then**   ▷ If any available traffic set for (m,k) is available
7:         $(m_i, k_i) = (m_{i+1}, k_{i+1})$ ▷ Change traffic set for next loose (m,k)
8:       **else**   ▷ If any available traffic set for (m,k) is not available
9:         $(m_i, k_i) = (m_s, k_s)$ ▷ Set the traffic as the least (m,k) requirement
10:         $payload\_size \leftarrow payload\_size \times \frac{1}{k}$ ▷ Traffic shaping by reducing packet size
11:         Payload_Size_Changed == TURE   ▷ Set the flag
12:       **end if**
13:     **else**   ▷ if only one DBP values are negative
14:       $(m_i, k_i) = (m_{i-1}, k_{i-1})$ ▷ Set the traffic as the next strict (m,k) requirement
15:     **end if**
16:   **else**   ▷ If the second DBP value become postive
17:     **if** $DBPT_i^{t+1} > 0$ **then**
18:       **if** $e == s$ && Packet_Size_Changed == TURE **then**   ▷ If the current set index is the largest value
19:         $payload\_size \leftarrow payload\_size \times k$ ▷ Restore the packet size
20:       **end if**
21:     **else**
22:       $(m_i, k_i) = (m_{i-1}, k_{i-1})$ ▷ Set the traffic as the next strict (m,k) requirement
23:     **end if**
24:   **end if**

- If the two values are negative as shown in lines 3 and 4, the current requirements of (m_i, k_i) are relaxed to (m_{i+1}, k_{i+1}), where the current index does not reach the maximum value. This is to allow more missed packets in S_i by loosing (m,k)-firm requirement. This kind of adjustment is available when any subset is available. Otherwise, if the current index is the maximum, another approach to reduce the payload size is applied.

- If the second DBP value becomes positive, the index for (m_i, k_i) increases to make current (m,k)-firm requirement strict because packet delay is not passed beyond deadline. This is, recovery procedure is invoked according to current positive DBP value.
- If the two consecutive DBP values are positive, two different tasks are performed according to the current index according to current index value. If the current index reaches the maximum value and the packet size is changed, the previous payload size is restored. Otherwise, (m_i, k_i) is set to close to (m_1, k_1) by decreasing the index number.
- If the DBP value becomes positive every time, the algorithm for adaptive traffic handling is not executed because the (m_i, k_i)-firm requirement is met by adaptive QoS routing.

As for all algorithms, Figure 2 shows the time sequence diagram for the proposed scheme. At first, path setup procedure is initiated. After receiving reply from the sink node, a sensor node starts delivering packets along the selected path according to (m,k)-firm requirements. The timeliness of packet is measured in a sink node and reported to a sensor node. According to this DBP value, a node decides whether new path is demanded or traffic shaping is required. If a next node detects event, a packet is delivered to the previous node and new path is searched. A previous node delivers packet from the next node along the path between itself and sink node. After new path is established, packet is delivered from current node to sink.

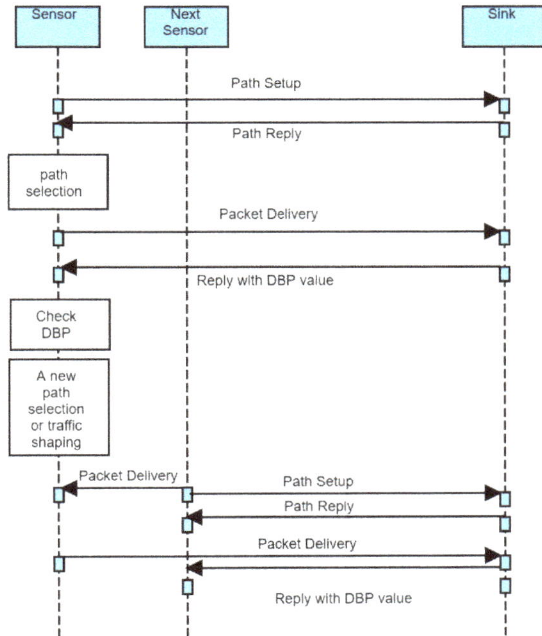

Figure 2. Overview of the proposed real-time model.

4. Performance Evaluation

This section presents the results of the simulation of the proposed scheme. We used the latest real-time routing protocol known as MK-AG [33] with the OPNET Modeler to simulate and analyze the proposed scheme. MK-AG selects a path based on multiple criteria, which are combined by the Analytical Hierarchical Process (AHP) in conjunction with the Gray Relational Analysis (GRA).

However, MK-AG considers the DBP value of the stream without regard to the requirements of the particular application.

Table 1 provides details of the values we used in our simulation. The nodes are placed in a terrain sized 1000×1000 m. One hundred nodes were used, of which 10 nodes were placed in the form of a grid, and the remaining 90 were randomly placed. Thus, the node density is determined arbitrarily. The results took the simulation environment into consideration. The transmission and receiving power consumption of a sensor node is 24.92 and 19.72 mJ, respectively, per byte. In order to generate traffic, we employ video traffic model in OPNET. A source node is selected randomly. A source node generate video traffic during random period. After period is over, next node in the source's transmission range is chosen. If the selected node cannot find the next source node, node is selected in backward direction continuously.

Table 1. Simulation environment setting.

Parameter	Value(s)
Terrain	(1000 m, 1000 m)
Number of nodes	100 Nodes
Node placement	Uniform & Random placement
Transmission range (m)	40 m
PHY and MAC protocol	802.15.4 PHY & MAC
Bandwidth	250 Kb/s
Payload size	64 bytes
Reduced payload size	32 bytes
Energy consumption (Tx)	24.92 mJ per 1 byte
Energy consumption (Rx)	19.72 mJ per 1 byte
Traffic model	Video traffic
Frame Interval Time	10 frame/s

The sink is located in the lower middle region of the field such that the end-to-end hop-count ranges from 4 to 9 hops with an average of 6 hops. The deadline for a real-time packet on each node was set to the average-link-delay $\times$ the number of shortest hops, respectively. We conduct the separate simulation to obtain average-link-delay. The evaluation result is presented as the stream dynamic success ratio (SDSR) based on stream dynamic failure ratio (SDFR) of the application. SDFR is the ratio sum of dynamic failures to total group of packets as much as k in the stream (total_number_of_packets/k). A dynamic failures happens if fewer than m out of any k consecutive packets are not delivered within their deadlines. We compute SDSR in following equation

$$SDSR = (1 - SDFR)$$
$$where$$
$$SDFR = \frac{sum\_of\_dynamic\_failures}{number\_of\_groups} \tag{3}$$

Additionally, we express the efficiency of the proposed scheme in terms of the network lifetime if the lifetime is prolonged.

For the (m,k)-firm stream for each S_i, the number of available stream sets, s, is set to 3. In addition, the value of t, which is used to monitor the DBP values, is set to 3. The parameter for reducing the packet size, k is set to 2 in Algorithm 2 such that a payload of 64 bytes is reduced to 32. As the background traffic, we employed three Constant Bit Rate (CBR) traffic streams and varied the amount of traffic to ensure the network is congested. Moreover, the source and destination of the background traffic was randomly chosen and was continuously changed after the predetermined duration. We calculate 95% confidence interval of average SDSR for 20 times evaluations with different 206 seed numbers for source and destination selection. Each simulation is conducted during 1000 s.

Simulation Results

The SDSR of the proposed scheme and that of the MK-AG routing protocol (denoted AG) are compared in Figures 3 and 4. Two respective (m,k)-firm streams were considered for application. Figure 3 shows the results of (2,3) as a more strict requirement than (2,5) in Figure 4. For each application, we compare three different (m,k)-firm streams. In other words, for each (m,k)-firm stream, three respective streams are composed. For example, the (3,4)-firm session stream consists of the set of {(3,4),(6,8), (9,12)}. Following the same approach, all sessions of the (m,k)-firm streams are defined as the set of three (m,k)-firm streams according to Equations (1) and (2).

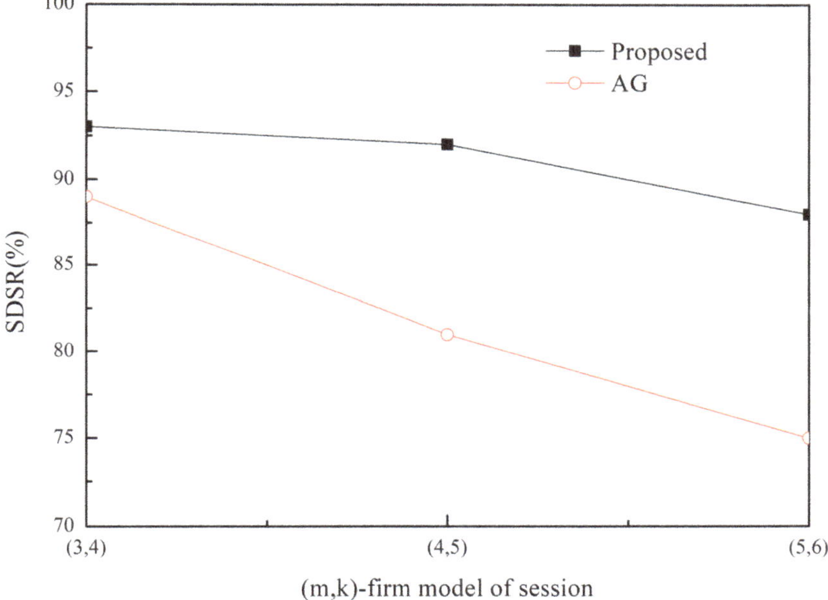

Figure 3. SDSR for (2,3)-firm application.

Figures 3 and 4 show that the SDSR of the strict requirement is lower than that of the less strict requirement. The same pattern is observed for the session requirements. That is, the highest SDSR is obtained for (3,6) session traffic in the (2,5)-firm application. In addition, the SDSR of the proposed scheme is higher than that of the MK-AG routing protocol. In Figure 3, the difference between the two schemes is larger where a strict requirement is demanded. The main reason for this gap is the adaptive routing protocol. Even though MK-AG takes into account the speed for real-time delivery, the delivery ratio on the link, and the remaining energy, it occasionally happens that a different path is not selected according to the DBP value. This implies that a different path is not selected when the multi-criteria approach is followed. In addition, the selection of a different path by the static comparison matrix in the AHP is time consuming. On the other hand, the proposed scheme makes use of multiple path selection according to the DBP value. Thus, because path selection is accomplished whenever the DBP value becomes negative, higher SDSR is achieved. The other important factor that affects the performance of the application is to take DBP_{ObjID} in DBP_{ObjID}. Because MK-AG only focuses on the session requirement, its compliance with the application requirements is less successful. However, the proposed scheme overcomes this problem by assigning a higher priority to DBP_{ObjID} in DBP_{ObjID} than DBP_i in S_i in Algorithm 1; hence, it promptly meets the requirement of the application according to DBP_{ObjID}.

In addition to using two schemes, adaptive traffic handling is performed according to the strict requirement. By using these diverse scenarios, we found that adaptive traffic handling was not invoked in too many cases, as would be expected for the strict (m,k)-firm requirement. More specifically, the adaptive traffic handling algorithm was executed 25 times in the (5,6)-firm session in the (2,3)-firm application, whereas this was not observed to occur in the (3,6)-firm session in the (2,5)-firm application. Among the 25 executions, five requests to change the packet size were detected. On the other hand, because the MK-AG routing protocol does not include an algorithm to adjust the traffic, more packets would be able to miss the deadline even though the shortest path is established. The effect of the traffic-handling algorithm is highlighted when strict requirements are used.

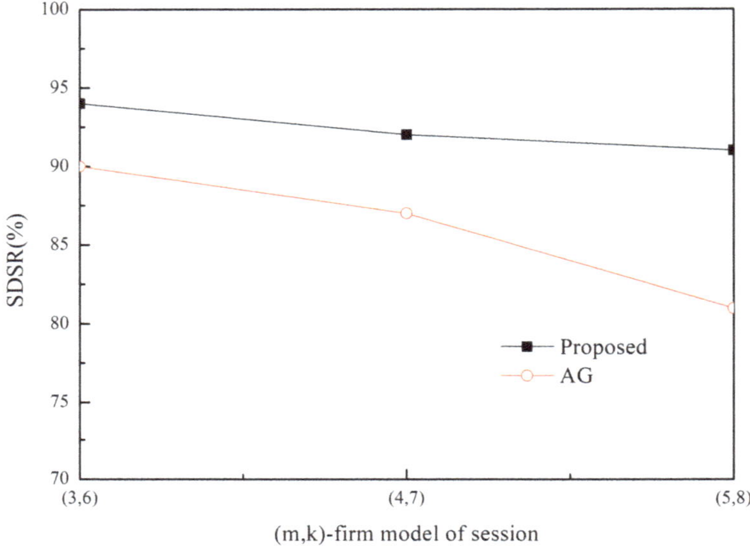

Figure 4. SDSR for (2,5)-firm application.

Usually control overhead may affect SDSR because high volume of control messages make network delay long. In the proposed scheme, control message includes RREQ and RREP during Path_Search_Probing(i,sink) procedure in Algorithm 1. Furthermore, a sink node periodically sends control message to a source node with DBP_{ObjID} and DBP_i. The former has negative effect in SDSR because it is based on the broadcast. However, its impact to SDSR is not great because the path search procedure is performed at the initial stage. Furthermore, control message between sink and source node is transmitted in a unicast way. Thus, its impact to SDSR is also limited. Similarly, control overhead of MK-AG is limited to path search and periodical reporting. In addition, there is no difference on the control overhead between streams because control message in the proposed scheme consists of path search and reporting. This implies that the performance gap between two protocols is observed by effectiveness of the adaptive path selection and traffic shaping. Specially, traffic shaping plays a great role in strict requirement by providing several alternative path and (m,k)-firm set.

We evaluated the performance when the network is congested by measuring the SDSR of the application as a function of the amount of background traffic. The results are plotted in Figures 5 and 6 for different (m,k)-firm sessions and applied schemes. For instance, 56AG represents (5,6)-firm stream session supported by MK-AG. The results of this evaluation exhibited a similar pattern as a function of different requirements. In all cases, the SDSR decreases as the amount of background traffic increases. The increased background traffic caused the network to become congested; therefore, more packets missed the deadline or were discarded. As compared to the MK-AG protocol, the proposed scheme is robust against congestion; consequently, adaptive path selection and traffic adjustment contribute to finding alternative paths. On the other hand, even though MK-AG takes congestion into account by considering the link delivery ratio, it does not compensate for lost packets as a result of congestion. Moreover, the static comparison matrix does not allow the source to change the path accordingly. As compared to light background traffic condition in Figures 3 and 4, the high impact of control overhead is observed in Figures 5 and 6. This is caused by the delay of control message at intermediate node as well as packet loss. These two features can lead to prevent establishing multiple paths and deliver the DBP values in early time. As a result, path selection and traffic shaping can be perform later than previous case, that is, light traffic condition. Thus, more packets missed deadline so lower SDSR is measured. This negative effect is observed in strict (m,k)-firm requirement more frequently.

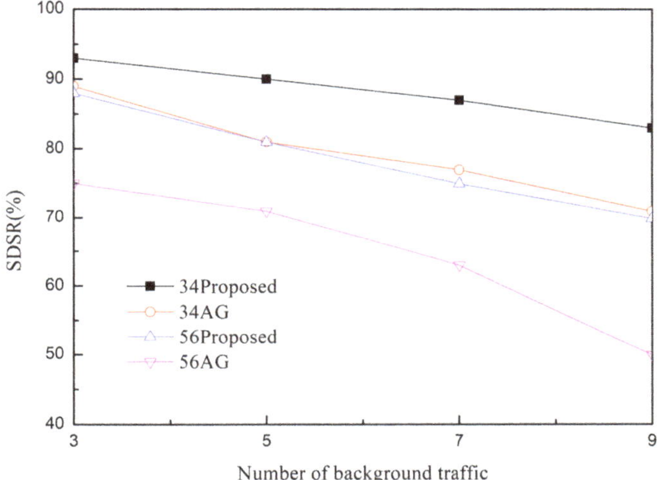

Figure 5. SDSR in (2,3)-firm application.

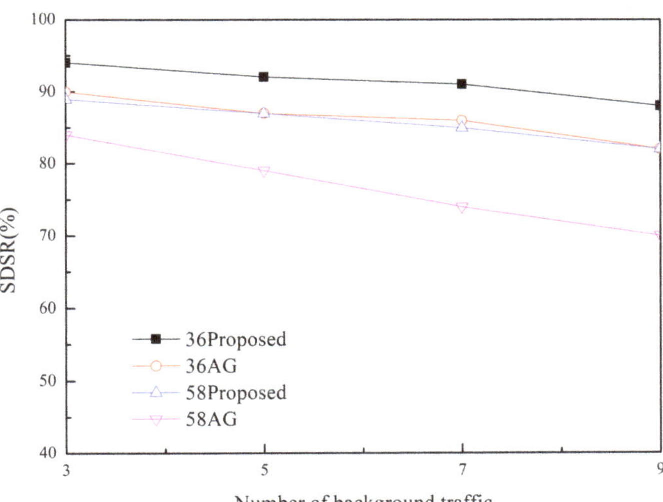

Figure 6. SDSR in (2,5)-firm application.

In addition to conducting a comparative study with the existing scheme, we conducted the simulation by varying the network environment and parameters in the proposed scheme.

Figure 7 shows the SDSR of the (2,3)-firm application as a function of the number of nodes. These results show that an increase in the number of nodes contributes to improving the SDSR. Because the node density is related to the number of paths, a large number of nodes enable more paths to be established between nodes. This means that the options of load balancing to prevent congestion and increased opportunities to change the path become available. Based on this analysis, great improvement is observed when using a more strict requirement such as a (5,6)-firm stream. The last simulation result in Figure 8 shows the impact of the number of (m,k)-firm sets on each session. We measured the SDSR by varying the number of sets from 2 to 4 in the adaptive traffic handling procedure in Algorithm 2. Higher SDSR is observed for a larger number of sets whereas the lowest values are measured for two sets. Because the larger number of levels provides a greater possibility to maintain a positive DBP value, enhanced SDSR values are obtained in all cases. In particular, a stricter (m,k)-firm streams benefits from a large number of sets.

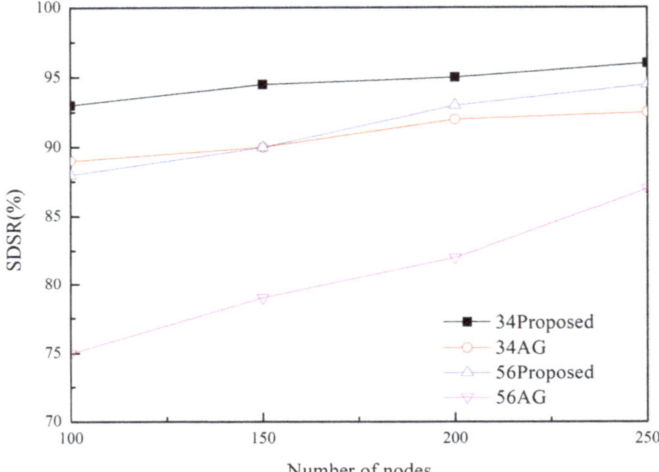

Figure 7. SDSR in (2,3)-firm application as a function of node density.

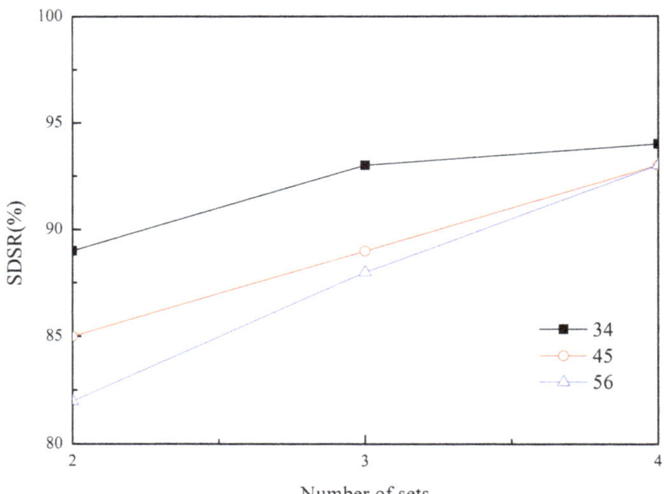

Figure 8. SDSR in (2,3)-firm application as a function of (m,k)-firm sets.

We measure the network lifetime of the proposed and existing scheme in two separate (*m,k*)-firm applications in Figures 9 and 10. In this paper, network lifetime is defined as elapsed time until the first sensor node or group of sensor nodes in the network runs out of energy. Because a node's energy consumption in the routing protocol is mostly dependent on the how many paths are established and how many packets are transmitted on a node, longer network lifetime than MK-AG is observed in the proposed scheme which is based on multiple paths which lead to distribute the packets along the multiple paths according to DBP values. In addition, adaptive path selection algorithm leads to prevent significant battery drain on nodes along the path. In the aspects of (*m,k*)-firm requirement, strict requirement is usually met by one or at most two paths. This limited number of path accelerates the battery drain on nodes. Therefore, short network lifetime is naturally observed in more strict requirement. In the other hand, network lifetime can be affected by the number of control message. Because our approach is based on path search through flooding, larger number of control messages lead to reduce network lifetime. In addition, reply message with DBP value is proportional to number of sent message. Despite of this shortcoming, longer network lifetime is achieved by the adaptive routing and traffic shaping. This implies that network lifetime is more dependent on the data packet than control one.

Finally, we conduct the simulation for SDSR for varying deadline as illustrated in Figures 11 and 12. In both figures, lower SDSR is observed shorter deadline is assumed in both schemes. However, higher SDSR is measured in the proposed scheme rather than MK-AG. Due to adaptive path selection and traffic adjustment algorithm lead to find the more suitable path than MK-AG. Even though larger gap between two schemes are measured in short deadline, almost identical SDSR is observed in the long deadline. This implies that two schemes are able to meet real-time delivery in appropriate way. In the aspects of (*m,k*)-firm requirement, less strict (*m,k*)-firm requirement induces higher SDSR than strict one. If the requirement is not strict, more available paths are likely to be established. This is mainly because packet loss and queuing delay caused by congestion are reduced.

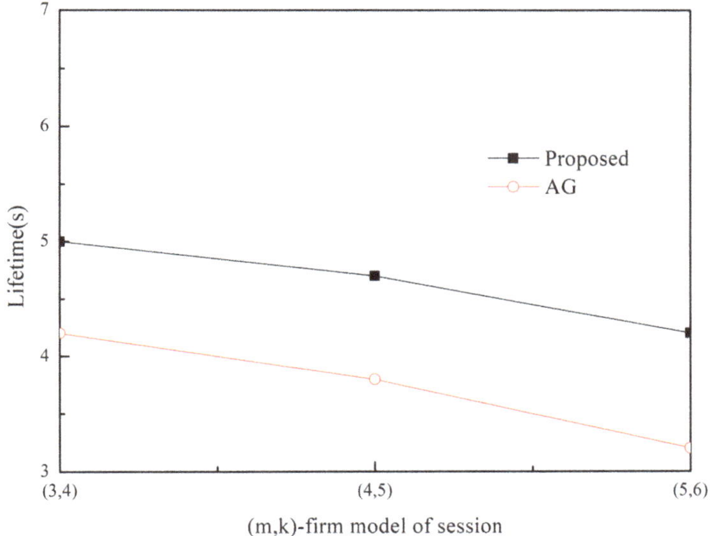

Figure 9. Lifetime in (2,3)-firm application as a function of (*m,k*)-firm sets.

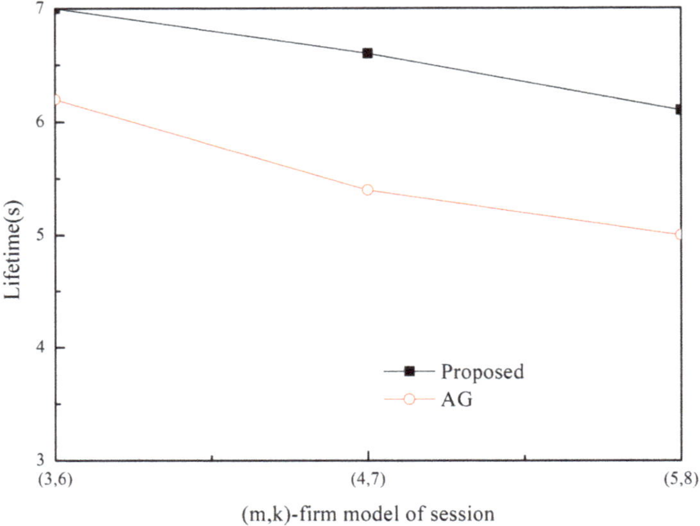

Figure 10. Lifetime in (2,5)-firm application as a function of (m,k)-firm sets.

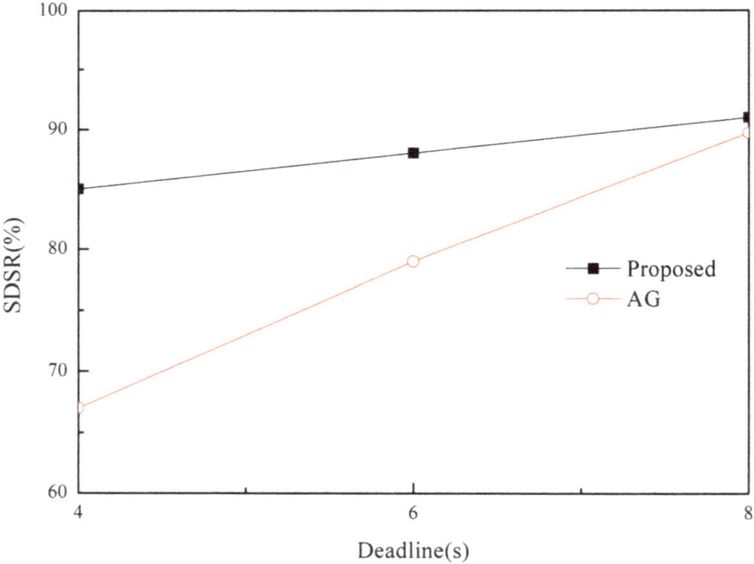

Figure 11. SDSR in (2,3)-firm application as a function of deadline.

As a result, we can summary the performance of the proposed scheme and MK-AG protocol in Table 2. We present the relative value according to (m,k)-firm requirement and given the metrics. As described in the Table 2, our proposed scheme is observed to provide more comparative performance than the existing scheme.

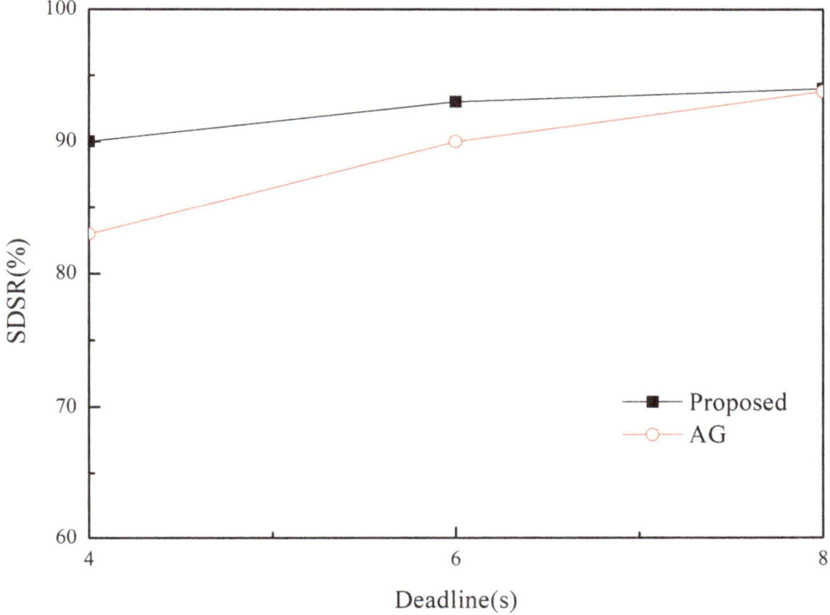

Figure 12. SDSR in (2,5)-firm application as a function of deadline.

Table 2. Summary for SDSR.

Parameters	(m,k)-Firm Type	Proposed	MK-AG
High Node Density	Strict	High	High
High Node Density	Loose	High	Medium
Low Node Density	Strict	High	High
Low Node Density	Loose	High	Low
Network Lifetime	Strict	Medium	Low
Network Lifetime	Loose	High	Medium
Short Deadline	Strict	Medium	Low
Short Deadline	Loose	High	Medium
Long Deadline	Strict	High	High
Long Deadline	Loose	High	High

5. Conclusions

In this paper, we presented a new real-time routing protocol and traffic handling algorithm for (m,k)-firm streams in wireless multimedia sensor networks. The application and session traffic were modeled by (m,k)-firm requirements and monitored by specifying a deadline for packet transmission. The current status of the (m,k)-firm requirement in both applications and sessions was used to choose an alternative path and adjust the traffic. Simulation results were provided to prove the efficiency of the proposed scheme compared to an existing scheme. The impact of the parameters was also analyzed.

Related to this scheme, we need to exploit real-time routing protocol to improve the energy efficiency of the proposed scheme. Furthermore, multimedia coding and compression scheme for (m,k)-firm streams will be studied. In addition, the improved adaptive path selection algorithm will be studied to reflect the latest end-to-end delay as well as predict next event. Finally, it will be required to introduce practical potential application such as object tracking to utilize the proposed mechanism.

Author Contributions: B.-S.K., S.K., K.H.K., T.-E.S., B.S., and K.-I.K. contributed to the design of the protocol and the detailed algorithm. They also conducted the performance analysis. All authors have read and agreed to the published version of the manuscript.

Funding: This research was supported by Basic Science Research Program through the National Research Foundation of Korea funded by the Ministry of Education (2018R1D1A1B07043731) and the Research Incentive Fund (R19045) of Zayed University, Abu Dhabi, UAE.

Conflicts of Interest: The authors declare no potential conflict of interests.

References

1. Hayat, M.; Khan, H.; Iqbal, Z.; Rahman, Z.; Tahir, M. Multimedia sensor networks: Recent trends, research challenges and future directions. In Proceedings of the International Conference on Communication, Computing and Digital Systems, Islamabad, Pakistan, 8–9 March 2017.
2. Gungor, V.; Hancke, G. Industrial Wireless Sensor Networks: Challenges, Design Principles, and Technical Approaches. *IEEE Trans. Ind. Electron.* **2009**, *56*, 4258–4265. [CrossRef]
3. Alanazi, A.; Elleithy, K. Real-Time QoS Routing Protocols in Wireless Multimedia Sensor Networks: Study and Analysis. *Sensors* **2015**, *15*, 22209–22233. [CrossRef] [PubMed]
4. Shen, H.; Bai, G. Routing in wireless multimedia sensor networks: A survey and challenges ahead. *J. Netw. Comput. Appl.* **2016**, *71*, 30–49. [CrossRef]
5. Bhandary, V.; Malik, A.; Kumar, S. Routing in Wireless Multimedia Sensor Networks: A Survey of Existing Protocols and Open Research Issues. *J. Eng.* **2016**, *2016*, 27. [CrossRef]
6. Hamdaoui, M.; Ramanathan, P. A dynamic priority assignment technique for streams with (m,k)-firm deadlines. *IEEE Trans. Comput.* **1995**, *44*, 1443–1451. [CrossRef]
7. Li, B.; Kim, K. An-firm real-time aware fault-tolerant mechanism in wireless sensor networks. *Int. J. Distrib. Sens. Netw.* **2012**, *20*, 719–731. [CrossRef]
8. Kafi, M.; Challal, Y.; Djenouri, D.; Doudou, M.; Bouabdallah, A.; Badache, N. A Study of Wireless Sensor Networks for Urban Traffic Monitoring: Applications and Architectures. *Procedia Comput. Sci.* **2013**, *19*, 617–626. [CrossRef]
9. Nellore, K.; Gerhard, P. A Survey on Urban Traffic Management System Using Wireless Sensor Networks. *Sensors* **2016**, *16*, 157. [CrossRef]
10. Moreno-Garcia, I.; Palacios-Garcia, E.; Pallares-Lopez, V.; Santiago, I.; Gonzalez-Redondo, M.; Varo-Martinez, M.; Real-Calvo, R. Real-Time Monitoring System for a Utility-Scale Photovoltaic Power Plant. *Sensors* **2016**, *16*, 770. [CrossRef]
11. Campos, A.; Garcia-Valdecasas, J.; Molina, R.; Castillo, C.; Alvarez-Fanjul, E.; Staneva, J. Addressing Long-Term Operational Risk Management in Port Docks under Climate Change Scenarios—A Spanish Case Study. *Water* **2019**, *11*, 2153. [CrossRef]
12. Huo, H.; Xu, Y.; Yan, H.; Mubeen, S.; Zhang, H. An Elderly Health Care System Using Wireless Sensor Networks at Home. In Proceedings of the Third International Conference on Sensor Technologies and Applications, Athens, Greece, 18–23 June 2009.
13. Velrani, K.S.; Geetha, G. Sensor based healthcare information system. In Proceedings of the IEEE Technological Innovations in ICT for Agriculture and Rural Development (TIAR), Chennai, India, 15–16 July 2016.
14. Dobslaw, F.; Zhang, T.; Gidlund, M. QoS-Aware Cross-Layer Configuration for Industrial Wireless Sensor Networks. *IEEE Trans. Ind. Inform.* **2009**, *12*, 1679–1691. [CrossRef]
15. Sun, W.; Yuan, X.; Wang, J.; Li, Q.; Chen, L.; Mu, D. End-to-End Data Delivery Reliability Model for Estimating and Optimizing the Link Quality of Industrial WSNs. *IEEE Trans. Autom. Sci. Eng.* **2018**, *15*, 1127–1137. [CrossRef]
16. Zhang, W.; Liu, Y.; Han, G.; Feng, Y.; Zhao, Y. An Energy Efficient and QoS Aware Routing Algorithm Based on Data Classification for Industrial Wireless Sensor Networks. *IEEE Access* **2018**, *6*, 46495–46504. [CrossRef]
17. Sarisaray-Bolukand, P.; Akkaya, K. Performance Comparison of Data Reduction Techniques for Wireless Multimedia Sensor Network Applications. *Int. J. Distrib. Sens. Netw.* **2015**, *11*, 873495. [CrossRef]
18. Bavarva, A.; Jani, P.; Ghetiya, K. Performance Improvement of Wireless Multimedia Sensor Networks Using MIMO and Compressive Sensing. *J. Commun. Inf. Netw.* **2018**, *3*, 84–90. [CrossRef]

19. He, T.; Stankovic, J.; Abdelzaher, T.; Lu, C. A spatiotemporal communication protocol for wireless sensor networks. *IEEE Trans. Parallel Distrib. Syst.* **2005**, *16*, 995–1006. [CrossRef]
20. Yu, Z.; Lu, B. A multipath routing protocol using congestion control in wireless multimedia sensor networks. *Peer Netw. Appl.* **2019**, *12*, 1585–1593. [CrossRef]
21. Ch, S.; Yim, Y.; Lee, J.; Park, H.; Kim, S. An opportunistic routing for real-time data in Wireless Sensor Networks. In Proceedings of the IEEE Wireless Communications and Networking Conference (WCNC), Shanghai, China, 7–10 April 2013.
22. Yang, T.; Kim, C.; Kim, S.; Lee, E.; Kim, S. Poster Abstract: Enhanced Real-Time Transmission Using Time Gain in Wireless Sensor Networks. In Proceedings of the 15th ACM/IEEE International Conference on Information Processing in Sensor Networks (IPSN), Vienna, Austria, 11–14 April 2016.
23. Hasan, M.; Al-Rizzo, H.; Al-Turjman, F. A Survey on Multipath Routing Protocols for QoS Assurances in Real-Time Wireless Multimedia Sensor Networks. *IEEE Commun. Surv. Tutor.* **2017**, *19*, 1424–1456. [CrossRef]
24. Park, J.; Jo, M.; Seong, D.; Yoo, J. Disjointed Multipath Routing for Real-Time Data in Wireless Multimedia Sensor Networks. *Int. J. Distrib. Sens. Netw.* **2014**, *10*, 783697. [CrossRef]
25. Deepaa, O.; Sugnuab, J. An optimized QoS-based clustering with multipath routing protocol for Wireless Sensor Networks. *J. King Saud. Univ. Comp. Inf. Sci.* **2017**, *9*. [CrossRef]
26. Kim, S.; Kim, B.; Kim, K.H.; Kim, K.I. Opportunistic Multipath Routing in Long-Hop Wireless Sensor Networks. *Sensors* **2019**, *19*, 4072. [CrossRef] [PubMed]
27. Kim, K.; Sung, T. Cross-Layered Approach for (m,k)-Firm Stream in Wireless Sensor Networks. *Wirel. Pers. Commun.* **2013**, *68*, 1883–1902. [CrossRef]
28. Li, B.; Kim, K. A novel real-time scheme for (m,k)-firm streams in wireless sensor networks. *Wirel. Netw.* **2014**, *20*, 719–731. [CrossRef]
29. Kim, B.; Park, H.; Kim, K.; Godfrey, D.; Kim, K. A Survey on Real-Time Communications in Wireless Sensor Networks. *Wirel. Commun. Mob. Comput.* **2017**, *2017*, 14. [CrossRef]
30. Lee, J.; Shah, B.; Pau, G.; Prieto, J.; Kim, K. Editorial: Real-Time Communication in Wireless Sensor Networks. *Wirel. Commun. Mob. Comput.* **2018**, *2018*, 2. [CrossRef]
31. Kim, K.; Sung, T. Modeling and Routing Scheme for (m,k)-firm Streams in Wireless Multimedia Sensor Networks. *Wirel. Commun. Mob. Comput.* **2015**, *15*, 475–483. [CrossRef]
32. Lee, C.; Shar, B.; Kim, K. An Architecture for (m,k)-firm Real-Time Streams in Wireless Sensor Networks. *Wirel. Netw.* **2016**, *22*, 69–81. [CrossRef]
33. Azim, M.; Kim, B.; Shah, B.; Kim, K. Real-time routing protocols for (m,k)-firm streams based on multi-criteria in wireless sensor networks. *Wirel. Netw.* **2017**, *23*, 1233–1248. [CrossRef]
34. Park, H.; Kim, B.; Kim, K.; Shah, B.; Kim, K. A Tree Based Broadcast Scheme for (m,k)-firm Real-Time Stream in Wireless Sensor Networks. *Sensors* **2017**, *17*, 2578. [CrossRef]
35. Kim, B.; Aldwairi, M.; Kim, K. An Efficient Real-Time Data Dissemination Multicast Protocol for Big Data in Wireless Sensor Networks. *J. Grid Comput.* **2019**, *17*, 341–355. [CrossRef]
36. Marina, M.; Das, S. Ad hoc on-demand multipath distance vector routing. *Wirel. Commun. Mob. Comput.* **2006**, *6*, 969–988. [CrossRef]

Article

A Spatial Group-Based Multi-User Full-Duplex OFDMA MAC Protocol for the Next-Generation WLAN

Meiping Peng, Bo Li, Zhongjiang Yan * and Mao Yang

School of Electronics and Information, Northwestern Polytechnical University, Xi'an 710072, China;
meiping@mail.nwpu.edu.cn (M.P.); libo.npu@nwpu.edu.cn (B.L.); yangmao@nwpu.edu.cn (M.Y.)
* Correspondence: zhjyan@nwpu.edu.cn; Tel.: +86-15877346761

Received: 16 May 2020; Accepted: 7 July 2020; Published: 9 July 2020

Abstract: The Wireless Local Area Network (WLAN) has become a dominant piece of technology to carry wireless traffic for Internet of Things (IoT). The next-generation high-density WLAN scenario is very suitable for the development trend of the industrial wireless sensor network. However, in the high-density deployed WLAN scenarios, the access efficiency is low due to severe collisions, and the interference is diffused due to the scattered locations of the parallel access stations (STAs), which results in low area throughput, i.e., low spatial reuse gain. A spatial group-based multi-user full-duplex orthogonal frequency division multiple access (OFDMA) (GFDO) multiple access control (MAC) protocol is proposed. Firstly, the STAs in the network are divided into several spatial groups according to the neighbor channel sensing ability. Secondly, a two-level buffer state report (BSR) information collection mechanism based on P-probability is designed. Initially, intra-group STAs report their BSR information to the group header using low transmission power. After that, group headers report both their BSR information collected from their members and inter-group interference information to the access point (AP). Finally, AP schedules two spatial groups without mutual interference to carry on multi-user full duplex transmission on the subchannels in cascading mode. The closed-form formulas are theoretically derived, including the number of uplink STAs successfully collected by AP, the network throughput and area throughput under saturated traffic. The simulation results show that the theoretical analysis coincide with the simulation results. The system throughput of the GFDO protocol is 16.8% higher than that of EnFD-OMAX protocol.

Keywords: WLAN; MAC; full duplex; spatial group; OFDMA; IEEE 802.11be

1. Introduction

As the main carrier of Internet of Things (IoT), wireless local area network (WLAN) is widely used due to its advantages of low cost and easy deployment [1,2] and has become a research hotspot in industry and academia. During the past decade, with the rapid development of mobile Internet, many new applications and requirements have emerged and people's demand for wireless traffic has increased rapidly at a compound annual growth rate of 47% from 2016 to 2021 [3], where the requirements for transmission delay and jitter are more stringent.

In order to achieve extremely high throughput, IEEE 802.11be [4] was established formally in May 2019 by the international organization of Institute of Electric and Electronics Engineering (IEEE), with the objectives to better support virtual reality, augmented reality, 4 K/8 K ultra-high definition video, remote office, and cloud computing application scenarios [5]. IEEE 802.11be continues to optimize orthogonal frequency division multiple access (OFDMA) technology and multi-user multiple input multiple output (MU-MIMO) technology introduced by IEEE 802.11ax, and takes multiple access points (multi-AP) cooperative and multi-band operation (MB-Opr) as its main technologies [6].

OFDMA technology divides channel resources into several subcarriers, and multiple successive subcarriers form a resource unit (RU). OMAX [7] is the first to propose a trigger free uplink multi-user random-access multiple access control (MAC) protocol based on OFDMA, so as to realize multi-user parallel data transmission.

However, there are still many technical bottlenecks to be solved when WLAN technology is directly applied to IoT scenarios, due to wireless scalability problems. Firstly, the network density scalability problem will appear if more and more sensors are deployed within a given network area. In other words, the low access efficiency problem of the high-density network will appear due to collisions between a large number of sensors/network devices, even if the multi-user MAC (MU-MAC) protocol, e.g., OFDMA, is employed. Secondly, the network deployment area scalability problem will appear if the sensors are deployed in a multi-hop mode over a large network area. In other words, the interference diffusion problem of the MU-MAC protocol in a high-density network will appear, where many stations (STAs)/sensors located at different positions will simultaneously transmit to AP/sink. It is harmful to exploit the spatial reuse gain. These two observations motivate us to solve the low access efficiency problem and the interference diffusion problem in high-density deployed WLAN, by exploiting the spatial grouping technology, the power control technology and full-duplex technology.

Based on the spatial grouping technology, Ref. [8] divides several spatial groups according to the spatial distance, reduces the interference diffusion problem of multi-user transmission, and improves the network area throughput. The employed spatial grouping technology in [8] can be combined with the power control technology, such that multiple spatial groups can be simultaneously transmitting or receiving, i.e., improve the spatial reuse gain. The power control technology can reduce the signal coverage, and thus eliminates the overlapping area and reduces the collisions [9–13]. In [14], it is proved that power control can effectively reduce the adjacent interference and improve the saturated throughput. Furthermore, if the adjacent interference is controlled and reduced by employing the power control technology, the Co-frequency Co-time Full Duplex (CCFD) technology can be exploited to enhance the overall network performance. Full duplex technology can transmit at the same time at the same frequency by using self-interference cancellation (SIC) technology [15,16], and thus the spectrum efficiency can be doubled. Therefore, it is regarded as the key piece of technology of the next generation of wireless communication [17,18].

The main contributions of this paper are summarized as follows:

1. A spatial Group-based multi-user Full Duplex OFDMA (GFDO) MAC protocol is proposed in this paper. To the best knowledge of the authors, GFDO is the first to jointly solve both the low-access efficiency problem and the interference diffusion problem in high-density deployed WLAN. This work is helpful to solve the wireless scalability problem in industrial wireless sensor networks. GFDO protocol is triggered by AP to collect uplink buffer state report (BSR) information and adopt a two-level BSR information collection mechanism to improve BSR collection efficiency. Meanwhile, in the second-level BSR collection process, the dynamic adjustment of the interference information between group headers is performed by detecting the power intensity, which improves the probability of forming full duplex link transmission successfully. Finally, AP schedules multi-user full duplex data transmission within each space group in a cascade manner according to the BSR information and group head interference information collected, the system throughput is greatly improved.

2. Through theoretical analysis of the proposed protocol, the closed expressions of the average nodes number of access channel, system saturation throughput and area throughput are derived.

3. The performance of the proposed GFDO protocol is compared with FD-OMAX protocol, EnFD-OMAX protocol, Mu-FuPlex protocol and OMAX protocol. The simulation results show that the theoretical analysis coincide with the simulation results, and the MAC efficiency of the proposed GFDO protocol is 16.8% higher than that of the EnFD-OMAX protocol.

The rest of this paper is organized as follows. In Section 2, the related works on full duplex MAC protocol are surveyed and the existing problems are analyzed. In Section 3, we describe the system model of GFDO protocol, including the target scenario, channel model, etc. In Section 4, we describe the process of GFDO protocol in details. In Section 5, the performance of the proposed GFDO protocol model is theoretically analyzed. Section 6 evaluates the performance of the proposed GFDO protocol, by comparing with other existing works through simulation results. This paper is summarized in Section 7.

2. Related Work and Motivation

According to the assumption of the network device full duplex capability, the existing MAC protocols can be divided into symmetrical and asymmetric MAC protocols. In the symmetric MAC protocols, all of the STAs have the full duplex capability. While in the asymmetric MAC protocols, only AP is assumed having full duplex capability. Therefore, we survey the existing related works according to the above two categories.

Symmetric full duplex MAC protocols usually assume that both STAs and AP have full duplex transmission capabilities. In other words, there can exists uplink and downlink transmissions simultaneously between one STA and one AP. In [19], a full duplex MAC protocol using frequency domain coordination was proposed for the next-generation WLAN. In the protocol, the AP specifies STAs to report channel information on the specified subchannel, and schedules full duplex link transmission, and STAs must have full duplex capability. In [20], a Carrier Sense Multiple Access/Collision Detect (CSMA/CD) like based multi-user full duplex MAC protocol with random competition in subchannels is proposed by using MAC frame preamble detection technology, which requires all nodes to have full duplex capability. That is, any node with data transmission request in the network sends data to the destination node as long as there are available subchannels, and the destination node transmits data to the source node at the same time.

Asymmetric full duplex MAC protocol means that AP has full duplex capability, but STAs are half duplex devices. Then, there could simultaneously exist one uplink transmission from one STA to AP, say STA A, and one downlink transmission from AP to another STA, say STA B. The communication links between STA A, AP and STA B are named as an asymmetric full duplex transmission links. With the next generation of WLAN devices miniaturization and low complexity, asymmetric full duplex MAC protocol will become a hotspot for the next-generation WLAN MAC protocol research. Q. Qu [21] et al. believed that in the next-generation WLAN full-duplex scenario, only the AP should have full duplex capability and STAS could not. Then, based on this assumption, the Mu-FuPlex [22] protocol and the PCMu-FuPlex [23] protocol based on AP pure scheduling are presented. However, the multi-user MAC protocol based on pure scheduling leads to high overhead in the process of BSR information collection. Therefore, the authors' previous studies have proposed a multi-user full duplex MAC protocol based on multi-user random access for the next-generation high-density deployment scenarios in FD-OMAX protocol [24], EnFD-OMAX protocol [25], which greatly improves the system throughput.

In summary, full duplex MAC protocol can increase system throughput, and Table 1 compares these MAC protocols from the following aspects, in terms of topology, contention based, performance metric, key features and evaluation method. However, all of the existing works do not take into account the collection efficiency of the uplink transmission requirements, and the STAs' location dispersion of multi-user transmission leads to serious interference diffusion. On the one hand, if the collection efficiency of uplink transmission requirements is low, the probability of forming a full duplex transmission pair is low, which directly leads to the low efficiency of full duplex MAC protocol. On the other hand, if the STAs positions involved in full duplex transmission are scattered, the overall interference area expands, which directly affects the area throughput of the system. Therefore, the access efficiency and interference diffusion are significant problems to be addressed, in the design of full duplex MAC protocol for the next-generation high-density WLAN.

Table 1. Comparison of MAC protocol for Co-time Co-frequency Full Duplex (CCFD) systems.

Reference	Type	Topology	Contention Based	Performance Metric	Key Features	Evaluation
FDC [19]	Symmetric	Centralized	Rotation	Throughput	Scheduling BSR collection on subchannels	Simulator
FD-CSMA/CD [20]	Symmetric	Distributed	Random access	Throughput	Random contention access on subchannel VMAC-hdr	Simulator
FuPlex [21]	Asymmetric	Distributed	RTS/CTS handshaking	Throughput Delay	Next-generation WLAN full duplex MAC framework	NS2 network simulator
Mu-FuPlex [22]	Asymmetric	Centralized	UORA	Throughput MAC Efficiency	Multi-users OFDMA	NS2 network simulator
FD-OMAX [24]	Asymmetric	Distributed	Random access	Throughput FD Link Efficiency	Multi-users OFDMA Inter-node interference collection	NS2 network simulator
EnFD-OMAX [25]	Asymmetric	Distributed	Random access	Throughput FD Link Efficiency MAC Efficiency	Multi-users OFDMA Inter-node interference collection full-duplex link pair matching algorithm	NS2 network simulator
PCMu-FuPlex [23]	Asymmetric	Centralized	UORA	Throughput	Multi-users OFDMA Power Control	NS2 network simulator
GFDO	Asymmetric	Centralized	*P*-probability	Throughput Access Efficiency Area Throughput	Multi-users OFDMA Spatial Group parallel BSR collection, Inter-node interference collection	NS2 network simulator

Based on our overview of the MAC protocol for CCFD transmission, this paper assumes that STAs can detect the received power intensity on the subchannel. Combining power control and spatial grouping technology, a spatial group-based multi-user full duplex OFDMA MAC protocol is proposed. Firstly, the AP divides the STAs into several spatial groups according to the neighbor channel sensing capability (NCSC) [26]. In spatial group, the STAs use the power control technology to report the BSR information to the group header in the low power mode. Secondly, group headers report the BSR information collected in this round and its own BSR information to the AP. In this process, the interference information among the group headers is dynamically updated. Finally, AP schedules multi-user full duplex transmission in the spatial group according to the interference information between spatial groups. The GFDO protocol uses trigger frames to manage the uplink/downlink transmission, which is compatible with the draft standard of IEEE-802.11ax/be.

3. System Model

This section mainly introduces the system model of GFDO protocol proposed in this paper. Before that, in order to better describe the system model, we firstly present the following definitions.

Definition 1: *Space Group (SG) is composed of STAs with the same NCSC. In a high-density deployed scenario, there are multiple SGs in a single basic service set (BSS) designated by AP.*

Definition 2: *Group Header (GH), designated by AP, is responsible for collecting BSR information of group members (GM) in a SG and interference information between GHs, and then reporting BSR information and interference information to AP.*

Definition 3: *Group Members (GM) are designated by AP according to the NCSC. The STAs with the same NCSC form a SG.*

Definition 4: *Group Full Duplex Transmission (GFDT): AP schedules GM in the uplink SG and in the downlink SG to form a GFDT according to the interference information between the GHs reported by GHs. As shown in Figure 1, GH_0 , AP and GH_4 form a GFDT, i.e., a spatial group-based asymmetric full duplex transmission link.*

In the GFDO protocol, the single BSS scenario for the next-generation WLAN is considered, which is composed of an AP with full duplex capability and several STAs with half duplex capability. As shown in Figure 1, AP is deployed at the geometric center of BSS, GMs are randomly distributed in several spatial groups under the coverage of AP, and the set of STAs is denoted as $S = \{S_1, S_2 \cdots S_{N_s}\}$. N_s GMs are divided into G_s SGs, denoted as $G = \{G_1, G_2 \cdots G_{G_s}\}$. Each SG contains one GH and N_g GMs. Suppose the whole bandwidth of the WLAN system is divided into N_r resource unit (RU), which is denoted as $R = \{R_1, R_2 \cdots R_{N_r}\}$.

The multi-user full duplex MAC protocol designed in this paper belongs to the asymmetric full duplex MAC protocol, as shown in Figure 1. GM_{0-1} and GM_{0-2} in SG_0 have uplink transmission requirements, GM_{4-2} and GH_4 in SG_4 have downlink transmission requirements. AP allocates channel resources R_1 to GM_{0-1} and GM_{4-2}, and channel resources R_{N_r} to GM_{0-2} and GH_4 to form full duplex transmission. The uplink transmission requirements of GMs in a SG are collected by GH and reported to the AP. In our model, in order to maximize the system throughput, we need to achieve the following conditions: (1) the collection efficiency of uplink transmission demand; (2) the collection of interference intensity between SGs.

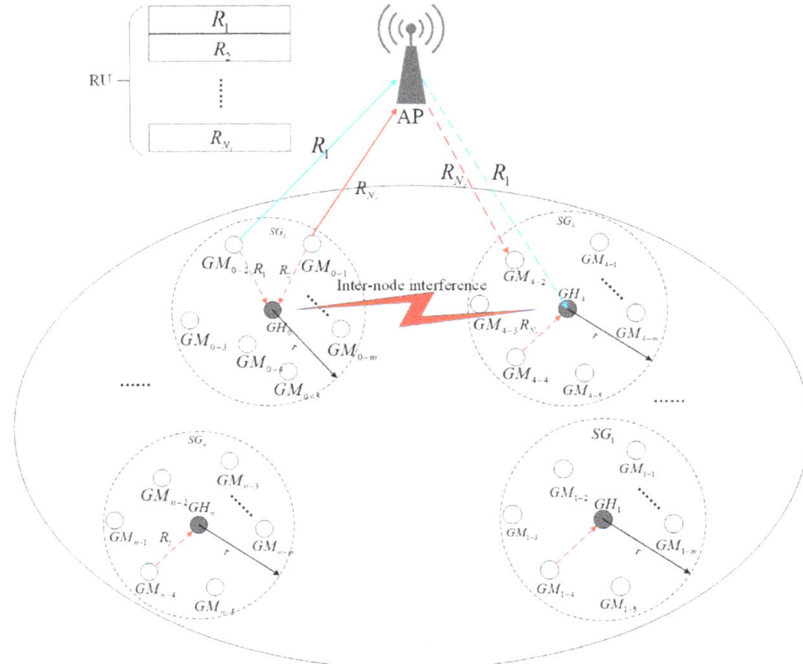

Figure 1. Full duplex communication network scenario under high-density deployment.

4. Protocol Description

4.1. The Basic Idea of GFDO Protocol

The GFDO protocol is an asymmetric full duplex MAC protocol for STAs random-access triggered by AP. The overall protocol flow is shown in Figure 2, including two stages, i.e., the two-level BSR information collection mechanism and inter-SGs interference collection, and the process of AP cascade scheduling GFDTs. Among them, the two-level BSR information collection mechanism is adopted for the uplink transmission demand collection. It includes the BSR information collection of GMs in SG and the BSR information collection of SG.

- In the first level, GH can be seen as a virtual gateway, which is used to collect the BSR information of GMs in SG and record the interference information of other GH. Because the formation of SG in GFDO protocol is based on the NCSC, the GMs in each SG can independently and synchronously report BSR information to the GH. Therefore, GFDO protocol can effectively improve the efficiency of BSR information collection, so as to improve the system throughput.
- In the second level, GHs report their GMs and their own BSR to AP if they have collected BSR in the first level. Otherwise, the GHs, with no collected BSR, dynamically update the inter-node interference intensity, which is almost real-time and improves the formation probability of a full duplex link.

It is worth noting that the compatibility of GFDO protocol with IEEE 802.11be is fully considered in the design, which has better backward compatibility.

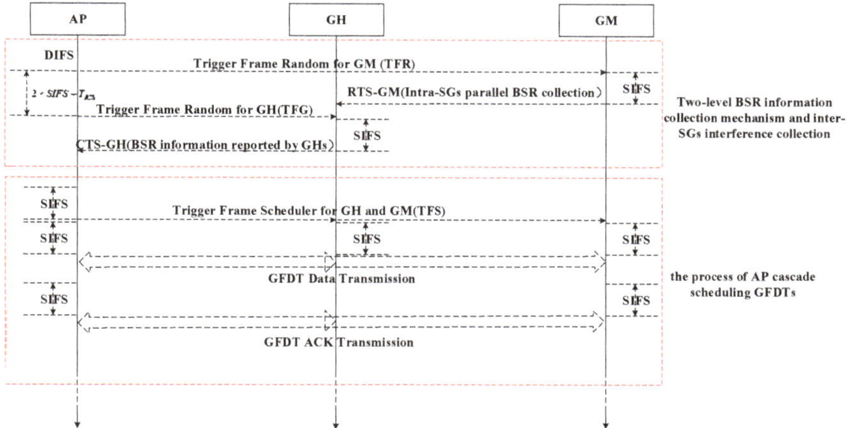

Figure 2. Overall flow chart of group-based multi-user full duplex orthogonal frequency division multiple access (OFDMA) (GFDO) protocol.

The key technologies of GFDO protocol are spatial grouping process and parallel collection of STAs uplink requirements in SG. First, we adopt the NCSC [26] to form the spatial groups. With NCSC, it means that if the distance between two nodes, say A and B, not exceed r (r refers to cell radius), then they will sense the channel activity status with the same probability. We define the receiving power of node A and node B at time t as $P_A(t)$ and $P_B(t)$, respectively, assuming that $P_d(t)$ is a difference of received power between receiving node A and receiving node B at time t. Then, the neighbor channel sensing ability can be expressed by Formula (1), where σ_d^2 is the variance of $P_d(t)$, σ_A^2 is the variance of $P_A(t)$ and σ_B^2 is the variance of $P_B(t)$. When ξ tends to 1, the nodes can accurately sense each other's channel state. That is, as long as the distance between nodes is close enough, their channel sensing ability is consistent. Through theoretical analysis and simulation verification, [26] proved the existence of neighbor channel awareness in high-density deployment scenarios and proved the correlation of the channel awareness ability of neighbor nodes. When the AP divides space groups, it refers to the analysis results of [26] and forms an SG with the same or similar NCSC. The implementation process is feasible.

$$\xi = 1 - \frac{\sigma_d^2}{\sigma_A^2 + \sigma_B^2} \tag{1}$$

As the core technology of IEEE 802.11 standard, power control has been put forward in the early standard version, e.g., IEEE 802.11 g. By adjusting the transmission power of AP and STAs, the collision and energy consumption can be reduced. The transmission power is reduced and the concurrent transmission rate in the network is increased [27]. Therefore, the spatial group BSR information collection proposed in this paper can achieve independent and parallel BSR collection process in each spatial group in the network by adjusting the transmission power of appropriate intra group BSR collection due to the same neighbor channel sensing capability of STAs in the group.

4.2. Protocol Process Description

The specific process of GFDO protocol is as follows.

- After AP successfully access into the channel, it starts the first level BSR information collection, i.e., GMs BSR information collection, by sending the BSR poll-trigger (BSRP-TFR) frame to start GMs uplink transmission demand collection in SG. Once the GMs receive BSRP-TFR frame, they report BSR to their respective GHs, with an power control based P-probability OFDMA random access method. In other words, once a GM receives BSRP-TFR, it randomly selects one RU and report its own BSR to its GH with the probability of P, and with a reduced transmission power. The reduced transmission power can only guarantee the BSR transmission to its own GHs, but with no harmful interference to other GHs.

- After the first-level BSR information collection, i.e., GMs BSR information collection, AP sends the BSRP-TFG frame to start the second-level BSR information collection. Once GHs receive the BSRP-TFG frame, they also adopt the P-probability OFDMA random access method to report two kinds of information to the AP. The first one is the BSR information collected in the first-level BSR information collection from the GMs, and its own BSR information if any. The second one is the interference information between SGs. However, if the GHs do not have any collected BSR information in the first-level BSR information collection, they will be in an idle state and do not report to AP. The idle GHs detect and record the interference intensity of other GHs on the subchannel.
- After two-level BSR information collection completion, the cascaded spatial group full duplex transmission is started. AP allocates RU resources according to the collected BSR and interference information among the GHs, and schedule a GFDT to start the multi-user full duplex transmission in cascaded mode.

4.2.1. Two-Level BSR Information Collection Mechanism

In the GFDO protocol, a two-level reporting mechanism is adopted for uplink BSR collection. The first-level BSR collection is that GHs as a virtual gateway independently and synchronously receive the BSR information reported by GMs in SG. The second-level AP collects the BSR information of GHs and the BSR information of GMs collected by GHs.

In the first level, the AP sends the BSRP-TFR frame to trigger all GMs to access the channel by OFDMA multiple access mode based on the joint control strategy of P-probability and power control. GHs are not allowed to compete with the channel and is in the receiving state. GMs use the same channel resource to report the BSR information to the corresponding GH independently and synchronously, so as to reduce the probability of access collision in the high-density deployment scenario. In addition, GMs randomly select an RU to send a request to send (RTS) frame after obtaining the transmission opportunity, which it does not send the RTS frame on the whole channel, thus further improving the probability of successfully accessing channel. As shown in Figure 1, all GMs in $G_0, \cdots, G_n$ are in the listening channel state. Once the BSRP-TFR frame sent by AP is listened to, after an short inter frame space (SIFS) time, all GMs randomly select a RU in range of $[0, RU_{N_r})$, and send the RTS frame to the respective GH based on P-probability on the selected RU. All of the GHs successfully receive the BSR frame sent by STAs in its SG and record the BSR information received. The BSR information set is recorded as follows:

$$GH_{bsr} = \left\{ BSR_{N_1}, BSR_{N_2} \cdots BSR_{N_g} \right\}$$

AP sets a timer after sending the BSRP-TFR frame, the length of which is the time of the first-level BSR information collection, defined as $\tau = t_{sifs} + t_{rts} + \Delta$, where t_{sifs} is SIFs time, t_{rts} is RTS frame transmission time, and Δ is signal transmission time. After the timeout of the timer, AP sends the BSRP-TFG frame after an SIFS time to trigger the second-level BSR information collection.

During the second-level BSR information collection, GMs are not allowed to access the channel, GHs use OFDMA multiple-access mode based on P-probability to compete for accessing the channel. Firstly, after receiving the BSRP-TFG frame, all of the GHs randomly selects an RU resource block within the range of $[0, RU_{N_r})$, and sends the group clear to send (G-CTS) frame on the selected RU. The G-CTS frame includes: BSR information collected at the first level, BSR information of the node itself and interference intensity information between SGs. As shown in Figure 1, the $GH_0, \cdots, GH_n$

report BSR information to the AP in the second-stage of BSR information collection. AP records all BSR information collected in this round, and its set is expressed as:

$$AP_{bsr} = \left\{ \begin{array}{l} \left\{ G_0, \left\{ (GM_{0,1}, bsr), (GM_{0,2}, bsr), \cdots, (GM_{0,N_g}, bsr) \right\} \right\}, \cdots, \\ \left\{ G_s, \left\{ (GM_{s,1}, bsr), (GM_{s,2}, bsr) \cdots (GM_{s,N_g}, bsr) \right\} \right\} \end{array} \right\}$$

The detailed flow of two-level BSR information collection mechanism is shown in Algorithm 1.

Algorithm 1: Two level BSR information reporting mechanism

1: **Global initialization:** //According to Equation (1), all of nodes in the network are divided into G_s space groups, each of which contains one GH and N_g GMs. Set the value of P-probability p.

2: **Step 1:** All GMs start the first level BSR information collection after receiving the BSRP-TFR frame sent by AP.

3:    a. All GMs pick two random value to prepare access channel and Check if data queue Q_{size} is not empty.
      $P = Random[0,1]$
      $RU_{index} = Random[0, RU_{N_r})$

4:    b. **IF** $P - p \geq 0$ and $Q_{size} > 0$ **THEN**

5:      GMs access channel on RU_{index} to send RTS frame in low power mode.

6:      **ELSE**

7:      GMs remain idle state.

8:      **ENDIF**

9:    c. GHs receive the RTS frame sent by GMs in the group, and records the received BSR information GH_{bsr}.

10. **Step 2:** After receiving the BSRP-TFG frame, GHs is ready to start the second level BSR information collection

11.   a. All GHs pick two random value to prepare access channel and Check if data queue Q_{size} is not empty.
      $P = Random(0,1]$
      $RU_{index} = Random[0, RU_{N_r})$

12. b. **IF** $(P - p \geq 0$ and $Q_{size} > 0)$ or $(P - p \geq 0$ and $GH_{bsr} > 0)$ **THEN**

13.   GHs access channel on RU_{index} to send G-CTS frame in full power mode.

14.   **ELSE**

15.   GHs keep receiving status and monitor the power intensity on RUs

16.   **ENDIF**

17.   c. The AP receives the G-CTS frame sent by GHs, and records all BSR information collected AP_{bsr}.

In summary, GFDO protocol adopts a two-level BSR information collection mechanism. On the one hand, AP divides nodes into several SGs according to NCSC, and specifies the transmission power of nodes in first level information collection. Therefore, when all of the GHs collect BSR information in respective SG simultaneously, there is no interference between each other, and the spatial reuse gain of the system channel resources is improved, so as to improve the area throughput of WLAN. On the other hand, OFDMA multiple-access mode based on P-probability strategy is adopted. Firstly, once a GM receives the trigger frame sent by AP, it randomly selects an RU within the range of the maximum resource block of the system. However, it does not immediately send the BSR information frame. The GMs that win the transmission opportunity based on P-probability strategy send the BSR information frame on the selected RU, so as to improve the node number of successful access channel, so as to further improve the system throughput of WLAN. Especially in the high-density network scenario for next-generation WLAN, massive amounts of nodes and APs are deployed in a limited area, the access collision probability is more serious. However, the GFDO protocol divides nodes into groups and competes for channels independently, which can significantly alleviate the access collision caused by a large number of nodes simultaneous access channels.

The collection of interference intensity between nodes is one of the problems to be solved in the design of full duplex MAC protocol. In order to improve the success probability of the full duplex transmission pair, the real-time inter-node interference intensity information as a key factor should be considered in full duplex link pairing. Therefore, the AP obtaining real-time inter-node interference intensity information is the cornerstone of improving the success probability of full duplex transmission.

In the second stage of BSR collection, GHs in an idle state monitor the channel and record the interference signal intensity of other GHs to this node. See Step 2 of Algorithm 1 for the collection and processing flow. As shown in Figure 1, assuming that GH_0 and GH_4 obtain access channel opportunities, G-CTS frames are sent in RU_3 and RU_1 respectively, while other GHs, such as GH_1 and GH_n, record the received power on different RU, and the set represents:$I_{GH} = \{(G_0, p), (G_1, p) \cdots (G_s, p)\}$, where p is the received power intensity of G-CTS frame on RU. Each time the new interference signal strength of the GH is detected, the corresponding interference intensity information is dynamically updated. In the second-level BSR information collection process, the GH writes the inter-node interference information into the G-CTS frame and reports it to the AP. After the AP receives the G-CTS frame sent by GH, it maintains a GH interference intensity information table, which set is:

$$I_{AP} = \left\{ \begin{array}{c} \{G_0, \{(G_1, p) \cdots (G_s, p)\}\}, \{G_1, \{(G_0, p) \cdots \\ (G_s, p)\}\} \cdots \{G_s, \{(G_0, p) \cdots (G_{s-1}, p)\}\} \end{array} \right\}$$

Because AP groups nodes according to NCSC, the GH and GMs in SG have the same channel sensing ability, that is, the interference intensity information of the GMs in the SG and the GMs in other SG can be represented by interference intensity information of the GH and other GHs.

4.2.2. Group Full Duplex Transmission in a Cascading Method

According to the description in the previous chapter, once the collection of uplink transmission demand and interference intensity information are complete, AP starts to schedule a GFDT in a cascading method. When the AP establishes a GFDT, it uses a Formula (2) to determine whether the power interference value transmitted by the uplink SG meets the signal to interference plus noise ratio (SINR) threshold received by the downlink SG, and form the pairing of the GFDT. Otherwise, a full duplex link cannot be formed. As shown in Figure 1, the power interference value of the uplink group G_0 meets the SINR threshold received by the downlink group G_4, which can form a GFDT.

$$SINR_{fd} = \frac{P_{r,DL(m)}}{N_0 + I_{DL(m)} + \Delta P} \tag{2}$$

where $P_{r,DL(m)}$ is the receiving power of downlink SG receiving downlink data on RU_m resource block. $I_{DL(m)}$ represents the interference power of uplink SG to downlink SG when it forms a full duplex transmission pair on the RU_m resource block. ΔP is the protection power value, this is, the average difference between GHs.

After AP completes the computation of the group full duplex transmission pairing, it sends the BSRP-TFC frame scheduling a GFDT. In the scheduling process, the SG is regarded as an entity, and Algorithm 2 shows the pseudo code of the GFDT scheduling in a cascading method, where the selection of the SINR threshold is based on [22]. AP schedules the group full duplex transmission according to the BSR information collected in this round. During the scheduling process, if the uplink demand collected by a group is greater than the number of RU in the system, the AP performs the cascade scheduling in this group until all nodes with uplink demands are collected in this group. If the number of uplink requirements collected by this group is less than the total number of RUs in the system, the out of group cascade scheduling will be performed until all uplink requirements collected are transmitted and the full duplex transmission of this group is completed. The unit of data transmission is the spatial group, and the information collected by BSR is taken as the benchmark, that is, scheduling members of two spatial groups that do not interfere with each other to form asymmetric group full duplex data transmission.

Algorithm 2: Group full duplex transmission in a cascading method

---

INPUT: Uplink transmission demand set AP_{bsr}, Downlink transmission demand set $DLset$, Inter-SGs interference intensity I_{AP}, SINR threshold $SINR_{throld}$
OUTPUT: Find $\left\{FD_n, S\left(GM_{i,ul}, GM_{j,dl}, RU_{index}\right)\right\} \forall GM_{i,ul} \in G_i, GM_{j,dl} \in G_j, i \neq j$, and $RU_{index} \in [0, RU_{N_r})$, FD_n is cascade number
1: **Initialization:**
2:    ULsize = size of AP_{bsr} // Number of SGs BSR information collected
3:    DLsize = size of $DLset$ // Number of Downlink SGs transmission demand
4:    $RU_{index} = 0$
5:    CasNum = 1
6: **FOR** ULsize **DO**
7: **FOR** DLsize **DO**
8:      Calculate the $SINR$ of GH_{ULsize} and GH_{DLsize} according to Equation (2) and Inter-SGs interference intensity I_{AP}
9:     **IF** $SINR > SINR_{throld}$ **THEN**
10:     in Size = size of $AP_{bsr}(ULsize)$ // Number of BSR information in a single SG
11:     **FOR** inSize **DO**
12:       Write full duplex link sets: $\left\{CasNum, S\left(GM_{i,inSize}, GM_{j,inSize}, RU_{index}\right)\right\}$
13:       $RU_{index}++$
14:       **IF** $RU_{index} \geq RU_{N_r}$ **THEN**
15:        CasNum++
16:        $RU_{index} = 0$
17:       **ENDIF**
18:      **ENDFOR**
19:     **ENDIF**
20:   **ENDFOR**
21: **ENDFOR**

4.2.3. Scheduling Strategy of Group Full Duplex Transmission

The GFDO protocol belongs to the AP centralized scheduling MAC protocol. According to the above sections, AP controls the BSR information and intergroup interference strength of the whole network node. In order to simplify the proposed protocol model, the fixed rate data transmission is adopted, so that the maximum full duplex transmission link pair can be formed in one transmission process is the optimization goal. Therefore, the bipartite graph matching algorithm is adopted to optimize the maximum full duplex link pair. Because only GHs participate in bipartite graph matching in GFDO protocol, and its computation complexity is $\Theta\left(n_{GH}^2\right)$, while all nodes participate in matching in other protocols, and its computation complexity is $\Theta\left[\left(n_{GH}\cdot N_g + n_{GH}\right)^2\right]$.

Because n_{GH}^2 is less than $\left(n_{GH}\cdot N_g + n_{GH}\right)^2$, the computation complexity of GFDO protocol is better than other protocols. Of course, the bipartite graph matching optimization algorithm is not the best method. It is our future research work to joint P-probability, User-AP Association power control and Modulation and Coding Schemes (MCS) to achieve the maximum throughput.

5. Performance Analysis

Based on the proposed GFDO protocol model, this section analyzes the average number of nodes in each round of the access channel and the system throughput under the ideal channel. During the analysis, we assume that the data queue of AP and STA is always non-empty and the network is saturated. Firstly, we define three variables used in the analysis of two-level BSR information collection: P-probability value p, the number of spatial groups G, and the number of nodes in the group N_g. Secondly, the two-level BSR information collection process is two independent collection processes.

Therefore, we analyze them separately. Finally, the closed expression of saturated throughput and the average number of STAs access channel are obtained, which provides effective theoretical support for the GFDO protocol proposed in the paper.

5.1. Analysis of the Average Number GMs of Access Channels in a Single SG

The success collection of BSR on RU depends on whether there is only one GM sending BSR request on the RU and no other GM sending BSR request on the RU. Let $P_{gm}(i)$ to access channel probability for i GM in each SG:

$$P_{gm}(i) = \begin{cases} 1 & if \quad p = 1 \\ \dbinom{N_g}{i} p^i (1-p)^{N_g-i} & if \quad p \in (0,1) \end{cases} \tag{3}$$

where N_g is the number of GMs. For each SG, there are i GMs sending a BSR request access channel in the current time slot, and each GM independently selects an RU. The probability of successful access channel is for each RU if and only if only one GM sends a BSR request on the RU, as shown in Equation (3):

$$P_{ru} = \frac{1}{M}(1 - \frac{1}{M})^{i-1} \tag{4}$$

The probability of GM sending BSR request on M RUs can be considered as M independent events with the same distribution, so the probability of average access channel in a competitive process is:

$$P_{m\text{-}ru} = MP_{ru} \tag{5}$$

Therefore, it can be concluded that the average number of GM of successful access channels in a single SG is:

$$N_{gm} = \sum_{i=1}^{N_g} iP_{m\text{-}ru}P_{gm}(i) \tag{6}$$

5.2. Analysis of the Average Number GHs of Access Channels in the System

In GFDO protocol, the way in which GMs compete for the access channel is the same as the GHs competing for the access channel. Define $P_{gh}(i)$ as the probability of preparing access channel for i GHs in the network:

$$P_{gh}(i) = \begin{cases} 1 & if \quad p = 1 \\ \dbinom{G}{i} p^i (1-p)^{G-i} & if \quad p \in (0,1) \end{cases} \tag{7}$$

As the GHs reports BSR information, M RU OFDMA mode is also used to access the channel. Similarly, the average number of GHs successfully accessed in a competition process is:

$$N_{gh} = \sum_{i=1}^{G} iP_{m\text{-}ru}P_{gh}(i) \tag{8}$$

According to the two-level information reporting mechanism proposed in this paper, that is, GMs successfully sends BSR request to the GH, and the GH successfully sends BSR request to AP, and the first-level transmission and the second-level transmission are independent events, the average number of BSR request successfully transmitted in the two-level information collection mechanism is:

$$N_{ap} = \sum_{i=1}^{G} iP_{m\text{-}ru}P_{gh}(i)\sum_{i=1}^{N_g} iP_{m\text{-}ru}P_{gm}(i) + \sum_{i=1}^{G} iP_{m\text{-}ru}P_{gh}(i) \tag{9}$$

5.3. Saturated throughput Analysis

According to the description in the previous chapter, the GFDO protocol can be regarded as a pure scheduling MAC protocol. In this section, based on the proposed GFDO protocol working in the ideal channel, the system throughput of the GFDO protocol is analyzed. The total length of a transmission can be divided into BSR information collection time and full duplex data transmission time, as shown in Figure 3. According to the description in Figure 3, define the time length of two stages, as shown in Equations (10) and (11).

$$T_{bsr} = T_{tfr} + T_{rts} + T_{tfg} + T_{cts} + 3 \cdot T_{sifs} + T_{difs} \tag{10}$$

$$T_{data}(i) = N_{gh} \cdot \left(T_{tfs} + T_{ack}(i) + T_{payload}(i) + 3 \cdot T_{sifs} \right) \tag{11}$$

where T_{tfr} is the transmission time length of BSRP-TFR frame, T_{rts} is the time length for STA to report BSR frame, T_{tfg} is the transmission time length of BSRP-TFG frame, T_{cts} is the time length of BSR frame reported by GH, T_{tfs} is the transmission time length of TFS frame, $T_{ack}(i)$ and $T_{payload}(i)$, respectively, represent the length of time when there are i pairs of full duplex links transmitting on different RU at the same time in the process of full duplex data transmission. Then, the expression of closed throughput in the saturated state is:

$$S = \begin{cases} \dfrac{E \sum\limits_{i=1}^{G} i P_{m\text{-}ru} P_{gh}(i) \left(\sum\limits_{i=1}^{N_g} i P_{m\text{-}ru} P_{gm}(i) \right)}{T_{bsr} + \sum\limits_{i=1}^{G} P_{m\text{-}ru} P_{gh}(i) \sum\limits_{i=1}^{N_g} P_{m\text{-}ru} P_{gm}(i) T_{data}(i)} & if \quad G = 1 \\[4ex] \dfrac{2E \sum\limits_{i=1}^{G} i P_{m\text{-}ru} P_{gh}(i) \left(\sum\limits_{i=1}^{N_g} i P_{m\text{-}ru} P_{gm}(i) \right)}{T_{bsr} + \sum\limits_{i=1}^{G} P_{m\text{-}ru} P_{gh}(i) \sum\limits_{i=1}^{N_g} P_{m\text{-}ru} P_{gm}(i) T_{data}(i)} & if \quad G \neq 1 \end{cases} \tag{12}$$

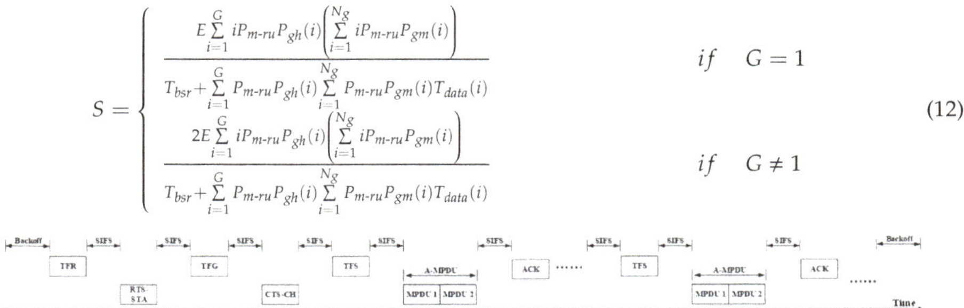

Figure 3. Transmission time mechanism of GFDO protocol.

5.4. Area Throughput Analysis

Area throughput refers to the amount of data sent by all nodes in the network in unit time and unit region, which is numerically equal to the ratio of the system throughput to the interference region of the sending node. In the paper, we assume that the interference region of the sending node is the region of the carrier sensing region when the node sending the frame, and the carrier sensing region radius is r_{cs}, and the NCSC radius is $r_{n\,csc}$, then the interference area of the group full duplex transmission can be expressed by Equation (13).

$$S_{n\,csc} = \pi (r_{cs} + r_{n\,csc})^2 \tag{13}$$

According to the throughput Equation (12), the area throughput can be represented by Equation (14).

$$S_{reg} = \begin{cases} \dfrac{E \sum\limits_{i=1}^{G} i P_{m\text{-}ru} P_{gh}(i) \left(\sum\limits_{i=1}^{N_g} i P_{m\text{-}ru} P_{gm}(i) \right)}{\left(T_{bsr} + \sum\limits_{i=1}^{G} P_{m\text{-}ru} P_{gh}(i) \sum\limits_{i=1}^{N_g} P_{m\text{-}ru} P_{gm}(i) T_{data}(i) \right) \cdot \pi (r_{cs} + r_{n\,csc})^2} & if \quad G = 1 \\[4ex] \dfrac{2E \sum\limits_{i=1}^{G} i P_{m\text{-}ru} P_{gh}(i) \left(\sum\limits_{i=1}^{N_g} i P_{m\text{-}ru} P_{gm}(i) \right)}{\left(T_{bsr} + \sum\limits_{i=1}^{G} P_{m\text{-}ru} P_{gh}(i) \sum\limits_{i=1}^{N_g} P_{m\text{-}ru} P_{gm}(i) T_{data}(i) \right) \cdot \pi (r_{cs} + r_{n\,csc})^2} & if \quad G \neq 1 \end{cases} \tag{14}$$

6. Performance Evaluation

6.1. Simulation Scene and Parameter Setting

In order to verify the efficiency of the GFDO protocol, a system and link level integrated simulation platform based on NS2 is built. In the simulation, we assume that in the process of receiving frames, as long as the received SINR is greater than or equal to the SINR threshold, the frame can be received successfully. In all simulations, the traffic flow is saturated, that is, each GM and GH always have uplink fame to AP and AP always has downlink frame to GM or GH. The time of each simulation is set to 50 s, and the average simulation result of 5 times of repeated simulation is finally taken. According to IEEE 802.11ax protocol draft, a 20 MHz full channel is divided into nine RUs, so, in the simulation scenario, nine STAs access channels at the same time are supported.

In the paper, we only consider the single BSS scenario. AP deployed in the geometric center of 100×100 m^2 simulation area, the number of GMs in each group N_g is 1, 3, 5, and the number of GH starts from 1, and increases gradually to 20 at one interval. The GHs were randomly distributed within the coverage of AP, and the GMs in SG were randomly distributed within the NCSC of the GH. According to the analysis of [26], the coverage of GH is set to 5 m. And the fixed QPSK modulation and 1/2 coding is used for data packet in all simulations. Other network parameter settings are shown in Table 2. The efficiency of BSR collection under different network scales, *P*-probability and number of GHs are verified. At the same time, the system performance of MU-FuPlex protocol and OMAX protocol is compared and verified, and its throughput is greatly improved. All the numerical simulation results verify the correctness of the proposed GFDO protocol.

BSR collection efficiency is the key performance metric in the GFDO protocol. BSR collection efficiency is a measure of system network capacity, which represents how many STAs participate in full duplex transmission at the same time. Therefore, the subsequent theoretical analysis and simulation verification in GFDO take BSR collection efficiency as the metric parameter to measure the system performance.

Table 2. Network parameter configuration.

Parameters	Value
Preamble Length	20 μs
PHY Rate	58.5 Mbps
RU Number	9
$SINR_{Throld}$	6 dB
DIFS	34 μs
SIFS	16 μs
Slot	9 μs
TXOP	0.003 s
Bandwidth	80 MHz

6.2. Simulation Result

6.2.1. Analysis of the Average Number of STAs Access Channel

The average number of STAs access network is an important performance metric to measure network performance. The more the average number of STAs access channel, the more uplink BSR information that AP collects, and the higher the scheduling efficiency of the GFDO protocol. The GFDO protocol model proposed combines the advantages of pure scheduling and random access, uses *P*-probabilistic in the BSR information collection stage and pure scheduling cascade transmission in the full duplex data transmission stage, which greatly improves the system performance. Figure 4 shows the average number of STAs access networks with a different *P*-probability. From Figure 4, we can see that the *P*-probability is 0.2, 0.4, 0.6, and 0.8, the number of GHs is 1–20, and the number of STAs in the SG is 1, 3, and 5 in Figure 4a–c, respectively. From the theoretical analysis and simulation curves in Figure 4a–c, the two curves are basically the same, which effectively verifies the correctness of the GFDO protocol. It provides reliable theoretical support for the proposed GFDO protocol.

As can be seen from Figure 4a, compared with the Mu-FuPlex and OMAX protocols, in the small-scale network scenario, due to the small competition conflict between the OMAX and Mu-FuPlex protocols, when the selected P-probability value is less than 0.8, the number of node access channels is better than the GFDO protocol. With the increase of the network scale, as shown in Figure 4a–c, the competition conflicts become more and more serious. As shown in Figure 4a–c, when $p = 0.6$ or 0.8, the number of access STAs of GFDO protocol based on the two-level BSR information collection mechanism is obviously better than that of Mu-FuPlex and OMAX protocol. The MU-FuPlex protocol adopts the competitive access mode of AP scheduling, and its average access channel number is less than that of OMAX.

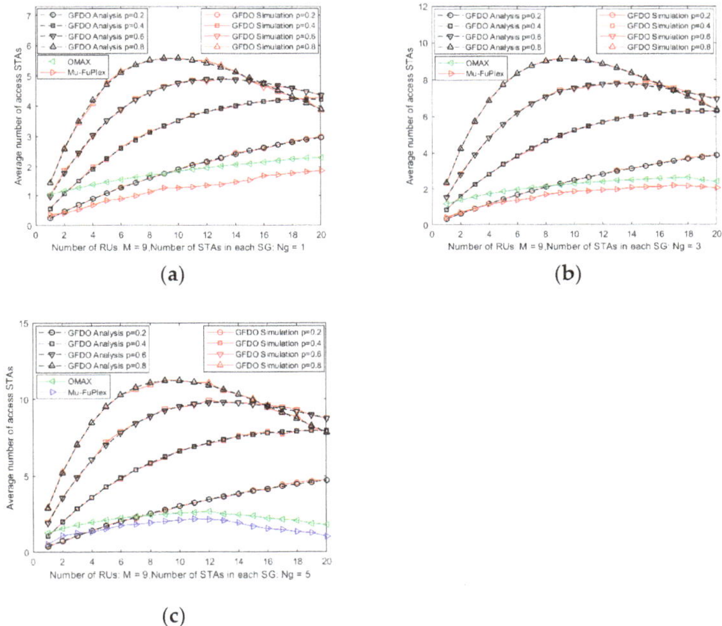

Figure 4. Analysis of the number of access STAs with a different P-probability. (**a**) The average number of access STAs vs. Number of STAs in each SG is 1; (**b**) The average number of access STAs vs. Number of STAs in each SG is 3; (**c**) The average number of access STAs vs. Number of STAs in each SG is 5.

6.2.2. Saturated Throughput Analysis

System throughput is an important performance metric to verify MAC protocol design. The purpose of the simulation is to verify the network throughput trend of GFDO protocol with the increase in node scale, and to compare with the theoretical analysis. From Figure 5, it can be seen that the simulation results are basically consistent with the theoretical analysis, further verifying the correctness of GFDO protocol, where the number of STAs in the SG is 1, 3, and 5 in Figure 5a–c, respectively.

As shown in Figure 5a–c, when the number of GHs is 1, all of the comparison protocols do not form full duplex link transmission, and the access efficiency of OMAX, EnFD-OMAX, FD-OMAX and Mu-FuPlex protocols is higher than the low P-probability access mode of GFDO protocol, and their throughputs are higher than GFDO protocol. However, with the increase in GHs and GMs, the randomness of competing nodes in EnFD-OMAX, FD-OMAX and Mu-FuPlex protocols reduces the probability of successfully forming full duplex link transmission, thus affecting the system throughput as shown in Figure 5b,c. The GFDO protocol divides the nodes into several SGs, and GMs in all of SGs compete independently and synchronously, which reduces the probability of competing channels at the same time. In addition, the GFDO protocol uses the information of GH interference to establish multi-user full duplex transmission in each SG, which improves the probability of successful full dual link transmission, thus greatly improving the system throughput. When the network scale reaches a certain number, as shown in Figure 5a, the throughput of OMAX, EnFD-OMAX, FD-OMAX, Mu-FuPlex protocols reduce sharply due to the aggravation of competition conflicts, while the GFDO protocol uses a two-level BSR information collection mechanism to divide the large-scale network into several small-scale networks, reducing competition conflicts, and the simulation results further verify that the GFDO protocol is applicable to the next-generation high-density and EHT WLAN scenarios.

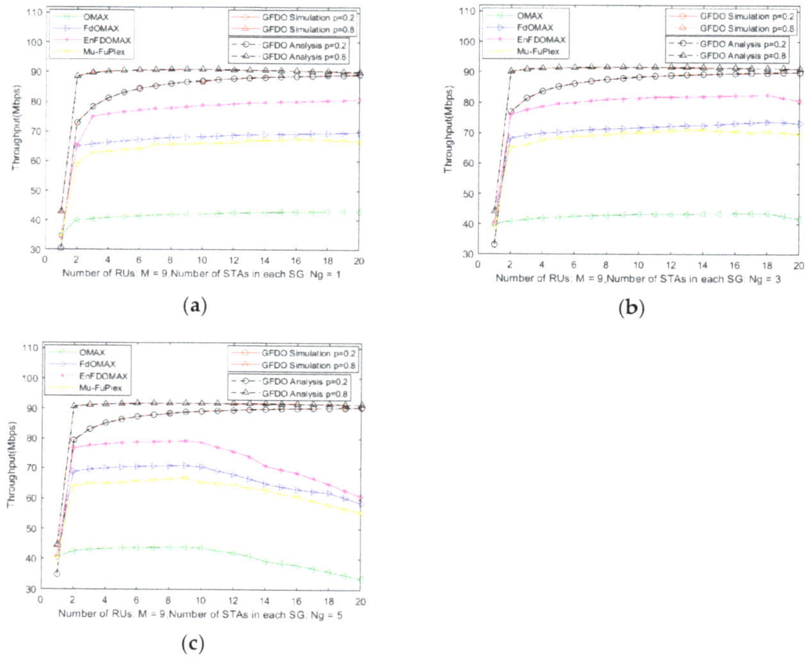

Figure 5. Analysis of system throughput under different *P*-probability. (**a**) Throughput vs. Number of STAs in each SG is 1; (**b**) Throughput vs. Number of STAs in each SG is 3; (**c**) Throughput vs. Number of STAs in each SG is 5.

6.2.3. Area Throughput Analysis

It can be seen from Figure 6 that the area throughput simulation results of GFDO protocol is consistent with the theoretical analysis, further verifying the correctness of GFDO protocol, where the number of STAs in the SG is 1, 3, and 5 in Figure 6a–c, respectively. When the number of nodes is small, the area throughput of all protocols is on the rise. From the simulation curve, it can be seen that the area throughput of GFDO protocol is improved due to other protocols, on the one hand, the access efficiency of GFDO protocol is improved, and the throughput is improved, which leads to the area throughput improvement. On the other hand, when forming a group full duplex link transmission, because the nodes are grouped according to NCSC, the users of parallel access are relatively concentrated, which reduces the interference diffusion problem, thus improving the area throughput.

As shown in Figure 6c, with the increase in node scale, the system throughput of other protocols decreases sharply due to the aggravation of competition conflicts, which leads to the decrease in area throughput. In the next-generation high-density deployment WLAN scenario, when the OMAX, FD-OMAX, EnFD-OMAX, Mu-FuPlex protocols perform multi-user parallel transmission, the location is relatively scattered, increasing the interference to the surrounding nodes, resulting in a sharp decline in area throughput.

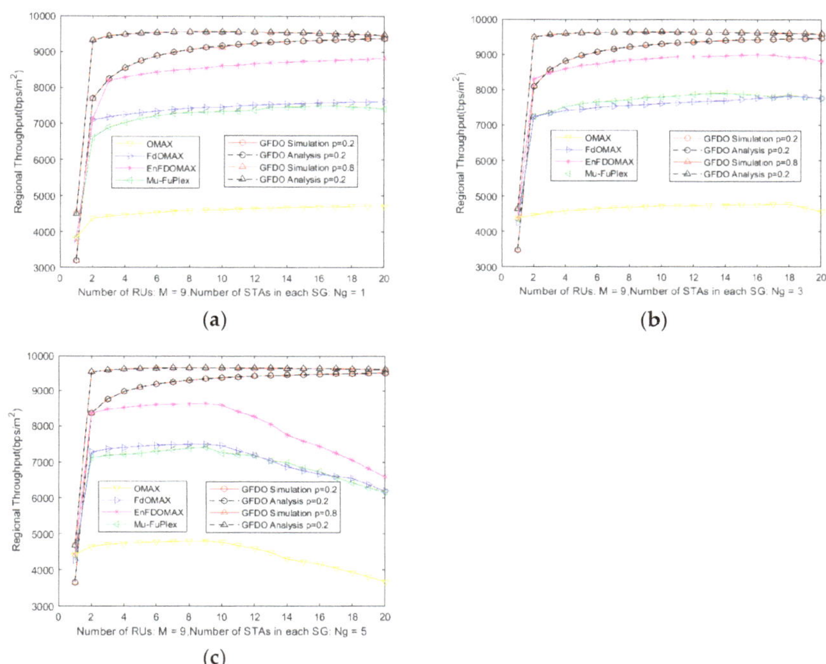

Figure 6. Analysis of area throughput under the number of different nodes. (**a**) Area throughput vs. Number of STAs in each SG is 1; (**b**) Area throughput vs. Number of STAs in each SG is 3; (**c**) Area throughput vs. Number of STAs in each SG is 5.

7. Conclusions

Aiming at the next-generation high-density deployed WLAN scenario, a spatial Group-based multi-user Full Duplex OFDMA (GFDO) MAC protocol was presented, to jointly solve both the low access efficiency problem and the interference diffusion problem. This protocol divides STAs into several spatial groups without interference, and obtains two advantages. Firstly, the high competition collision probability in high-density deployment scenario is reduced. Secondly, the formation probability of the full duplex transmission pair between the non-interference spatial groups is greatly improved. The simulation results show that the MAC efficiency of the proposed GFDO protocol is 16.8% higher than that of the EnFD-OMAX protocol. In the future, we will focus on the enhancement of GFDO protocol in the optimization algorithm of *P*-probability access and experiment to verify the validity of GFDO protocol.

Author Contributions: Conceptualization, M.P., B.L., Z.Y. and M.Y.; Data curation, M.P.; Formal analysis, Z.Y.; Funding acquisition, B.L., Z.Y. and M.Y.; Investigation, M.P. and Z.Y.; Methodology, M.P., Z.Y. and M.Y.; Project administration, B.L., Z.Y. and M.Y.; Resources, M.P.; Software, M.P.; Supervision, B.L.; Writing—original draft, M.P. Writing—review & editing, B.L., Z.Y. and M.Y. All authors have read and agreed to the published version of the manuscript.

Funding: This work was supported in part by the National Natural Science Foundations of CHINA (Grant No.61771392, 61771390, 61871322 and 61501373) and the Science and Technology on Avionics Integration Laboratory (Grant No. 20185553035 and 201955053002).

Conflicts of Interest: The authors declare no conflict of interest.

References

1. Pirayesh, H.; Sangdeh, P.K.; Zeng, H. EE-IoT: An Energy-Efficient IoT Communication Scheme for WLANs. In Proceedings of the IEEE INFOCOM 2019 -IEEE Conference on Computer Communications, Paris, France, 29 April–2 May 2019; pp. 361–369.

2. Teng, R.; Yano, K.; Kumagai, T. Scalable distributed-sensing scheme with prioritized reporting for multi-band WLANs. In Proceedings of the 2018 IEEE 4th World Forum on Internet of Things (WF-IoT), Singapore, 5–8 February 2018; pp. 580–585.

3. Cisco. *Cisco Visual Networking Index: Global Mobile Data Traffic Forecast Update; 2016c2021 White Paper; Tech. Rep. Cisco White Paper;* Cisco: San Jose, CA, USA, 2017.

4. IEEE P802.11-Task Group BE (EHT) Meeting Update. Available online: http://www.ieee802.org/11/Reports/tgbe_update.htm (accessed on 30 April 2020).

5. Lopez-Perez, D.; Garcia-Rodriguez, A.; Galati-Giordano, L.; Kasslin, M.; Doppler, K. IEEE 802.11be Extremely High Throughput: The Next Generation of Wi-Fi Technology beyond 802.11ax. *IEEE Commun. Mag.* **2019**, *57*, 113–119. [CrossRef]

6. Yang, M.; Li, B.; Yan, Z.; Yan, Y. AP Coordination and Full-duplex enabled Multi-band Operation for the Next Generation WLAN: IEEE 802.11be (EHT). In Proceedings of the 2019 11th International Conference on Wireless Communications and Signal Processing (WCSP), Xi'an, China, 23–25 October 2019; pp. 1–7.

7. Qu, Q.; Li, B.; Yang, M.; Yan, Z. An OFDMA based concurrent multiuser MAC for upcoming IEEE 802.11ax. In Proceedings of the 2015 IEEE Wireless Communications and Networking Conference Workshops (WCNCW), New Orleans, LA, USA, 9–12 March 2015; pp. 136–141.

8. Li, Y.; Li, B.; Yang, M.; Yan, Z. A spatial clustering group division-based OFDMA access protocol for the next generation WLAN. *Wirel. Netw.* **2019**, *25*, 5083–5097. [CrossRef]

9. Demirci, I.; Korcak, O. Gap-free load balancing algorithms in wireless LANs using cell breathing technique. In Proceedings of the 2014 IEEE International Conference on Advanced Networks and Telecommuncations Systems (ANTS), New Delhi, India, 14–17 December 2014; pp. 1–6.

10. Tam, H.H.M.; Tuan, H.D.; Ngo, D.T.; Duong, T.Q.; Poor, H.V. Joint Load Balancing and Interference Management for Small-Cell Heterogeneous Networks With Limited Backhaul Capacity. *IEEE Trans. Wirel. Commun.* **2017**, *16*, 872–884. [CrossRef]

11. Iwai, K.; Ohnuma, T.; Shigeno, H.; Tanaka, Y. Improving of Fairness by Dynamic Sensitivity Control and Transmission Power Control with Access Point Cooperation in Dense WLAN. In Proceedings of the 2019 16th IEEE Annual Consumer Communications & Networking Conference (CCNC), Las Vegas, NV, USA, 11–14 January 2019.

12. Han, L.; Zhang, Y.; Zhang, X.; Mu, J. Power Control for Full-Duplex D2D Communications Underlaying Cellular Networks. *IEEE Access* **2019**, *7*, 111858–111865. [CrossRef]

13. Choi, W.; Lim, H.; Sabharwal, A. Power-Controlled Medium Access Control Protocol for Full-Duplex WiFi Networks. *IEEE Trans. Wirel. Commun.* **2015**, *14*, 3601–3613. [CrossRef]

14. Liu, S.; Fu, L.; Xie, W. Hidden-Node Problem in Full-Duplex Enabled CSMA Networks. *IEEE Trans. Mob. Comput.* **2020**, *19*, 347–361. [CrossRef]

15. Han, S.; Zhang, Y.; Meng, W.; Chen, H. Self-Interference-Cancelation-Based SLNR Precoding Design for Full-Duplex Relay-Assisted System. *IEEE Trans. Veh. Technol.* **2018**, *67*, 8249–8262. [CrossRef]

16. Hong, S.; Mehlman, J.; Katti, S. Picasso: Flexible RF and Spectrum Slicing. *Comput. Commun. Rev.* **2012**, *42*, 283–284. [CrossRef]

17. IEEE802.11. HEW MAC Efficiency Analysis for HEW SG, IEEE 802. 11-13/ 0765r1. Available online: https://mentor.ieee.org/802.11/dcn/13/11-13-0765-02-0hew-co-time-co-frequency-full-duplex-for-802-11-wlan.ppt (accessed on 17 July 2013).

18. IEEE802.11. HEW MAC Efficiency Analysis for HEW SG. IEEE 802. 11/18-1225r0. Available online: https://mentor.ieee.org/802.11/dcn/18/11-18-1225-00-00fd-technical-report-on-full-duplex-for-802-11-fd-architecture.docx (accessed on 9 July 2018).

19. Ahn, H.; Lee, J.; Kim, C.; Suh, Y. Frequency Domain Coordination MAC Protocol for Full-Duplex Wireless Networks. *IEEE Commun. Lett.* **2019**, *23*, 518–521. [CrossRef]

20. Wang, X.; Tang, A.; Huang, P. Full duplex random access for multi-user OFDMA communication systems. *Ad Hoc Netw.* **2015**, *24*, 200–213. [CrossRef]

21. Qu, Q.; Li, B.; Yang, M.; Yan, Z.; Zuo, X.; Guan, Q. FuPlex: A full duplex MAC for the next generation WLAN. In Proceedings of the 2015 11th International Conference on Heterogeneous Networking for Quality, Reliability, Security and Robustness (QSHINE), Taipei, Taiwan, 19–20 August 2015; pp. 239–245.

22. Qu, Q.; Li, B.; Yang, M.; Yan, Z.; Zuo, X. MU-FuPlex: A Multiuser Full-Duplex MAC Protocol for the Next Generation Wireless Networks. In Proceedings of the 2017 IEEE Wireless Communications and Networking Conference (WCNC), San Francisco, CA, USA, 19–22 March 2017; pp. 1–6.

23. Qu, Q.; Li, B.; Yang, M.; Yan, Z. Power Control Based Multiuser Full-Duplex MAC Protocol for the Next Generation Wireless Networks. *Mob. Netw. Appl.* **2018**, *23*, 1008–1019. [CrossRef]

24. Peng, M.; Li, B.; Yan, Z.; Yang, M. Research on a trigger-free multi-user full-duplex MAC protocol. *Comput. Simul.* **2020**, accepted.

25. Peng, M.; Li, B.; Yan, Z.; Yang, M. A Trigger-Free Multi-user Full Duplex User-Pairing Optimizing MAC Protocol. *IoTaaS* 2019. In Proceedings of the 5rd EAI International Conference on IoT as a Service, Xi'an, China, 16–17 November 2019; pp. 598–610. [CrossRef]

26. Qu, Q.; Li, B.; Yang, M.; Yan, Z.; Zuo, X.; Zhang, Y. The neighbor channel sensing capability for wireless networks. In Proceedings of the 2016 IEEE International Conference on Signal Processing, Communications and Computing (ICSPCC), Hong Kong, China, 5–8 August 2016; pp. 1–6.

27. Cho, Y.; Kim, Y.; Lee, T. Energy-Perceptive MAC for Wireless Power and Information Transfer. *IEEE Wirel. Commun. Lett.* **2019**, *8*, 644–647. [CrossRef]

Article

Distributed Node Scheduling with Adjustable Weight Factor for Ad-hoc Networks

Wonseok Lee, Taehong Kim and Taejoon Kim *

School of Information and Communication Engineering, Chungbuk National University, Chungju 28644, Korea; wonseoklee@cbnu.ac.kr (W.L.); taehongkim@cbnu.ac.kr (T.K.)
* Correspondence: ktjcc@chungbuk.ac.kr

Received: 31 July 2020; Accepted: 5 September 2020; Published: 7 September 2020

Abstract: In this paper, a novel distributed scheduling scheme for an ad-hoc network is proposed. Specifically, the throughput and the delay of packets with different importance are flexibly adjusted by quantifying the importance as weight factors. In this scheme, each node is equipped with two queues, one for packets with high importance and the other for packets with low importance. The proposed scheduling scheme consists of two procedures: intra-node slot reallocation and inter-node reallocation. In the intra-node slot reallocation, self-fairness is adopted as a key metric, which is a composite of the quantified weight factors and traffic loads. This intra-node slot reallocation improves the throughput and the delay performance. Subsequently, through an inter-node reallocation algorithm adopted from LocalVoting (slot exchange among queues having the same importance), the fairness of traffics with the same importance is enhanced. Thorough simulations were conducted under various traffic load and weight factor settings. The simulation results show that the proposed algorithm can adjust packet delivery performance according to a predefined weight factor. Moreover, compared with conventional algorithms, the proposed algorithm achieves better performance in throughput and delay. The low average delay while attaining the high throughput ensures the excellent performance of the proposed algorithm.

Keywords: distributed scheduling; weight factor; ad-hoc network; fairness

1. Introduction

An ad-hoc network, as a group of wireless mobile nodes, can be implemented in various forms, including wireless mesh networks, wireless sensor networks, mobile ad-hoc networks, and vehicle ad-hoc networks [1,2]. Ad-hoc networks can provide flexible communication even when it is not possible to install new infrastructure or use existing infrastructure due to geographical and cost restrictions [3]. Ad-hoc networks have the advantage of node communication with other nodes without a base station. Moreover, they also have the features of self-forming and self-healing. Accordingly, they are adopted in various applications, such as battlefield situations, where topology changes frequently, disaster relief, environmental monitoring, smart space, medical systems, and robot exploration [4–8].

Unlike mobile communication networks, which allow centralized resource scheduling, an ad-hoc network requires distributed scheduling based on the information exchanged among nodes. A major problem with distributed node scheduling is packet collisions among nodes if resources are not efficiently distributed, which can lead to significant throughput degradation [9]. Considering these characteristics, supporting quality of service (QoS) through distributed scheduling is a very challenging task. QoS support for high- and low-priority data is essential in various applications. For instance, on a battlefield, a commander's orders must be delivered as soon as possible. In addition, for environmental monitoring, it is necessary to send emergency disaster information, such as an earthquake alert, to a destination node with very high priority [10].

The nodes of an ad-hoc network consume a lot of energy in sensing data and processing high-priority packet. However, in many situations, it is difficult to replace or recharge the battery of the nodes. Accordingly, it is important to increase energy efficiency and to enhance overall network lifetime through clustering, transmission power control, and efficient network information exchange [11–16]. Fairness and load balancing among nodes also have a great influence on the battery lifetime and the connectivity of the entire network. However, low fairness among nodes due to inefficient resource allocation causes increased packet collisions and packet retransmission to some nodes, and these detrimental effects reduce the battery lifetime. Meanwhile, some other nodes will be allocated an unnecessarily much amount of resources, resulting in severe inefficiency for the entire network. Hence, resource allocation for an ad-hoc network is a very important and challenging issue.

Fairness measurements can be categorized into qualitative and quantitative methods, depending on whether the fairness can be quantified. Qualitative methods cannot quantify fairness to an actual value, but they can judge whether a resource allocation algorithm achieves a fair allocation. Maximum-minimum fairness [17,18] and proportional fairness [19] are qualitative methods. Maximum-minimum fairness aims to achieve a max-min state, where the resources allocated to a node can no longer be increased without reducing the resources allocated to neighboring nodes. Proportional fair scheduling maximizes the log utility of the whole network by preferentially scheduling nodes with the highest ratios of currently achievable rates to long-term throughput. Measuring the fairness of an entire network is also an important issue. Jain's fairness index [20] is a quantitative fairness measurement method, however, it cannot measure the fairness of nodes to which a weight factor is assigned.

In this paper, a distributed scheduling algorithm, which takes weight factors and traffic load into account, is proposed. In the proposed algorithm, self-fairness [21] is adopted for resource reallocation. Increment of self-fairness means that resources are fairly allocated to nodes proportionally to the weight of each node. Therefore, even in the distributed scheduling which supports packets with different importance, if the slot allocation for each node is adjusted to the direction of increasing self-fairness, the overall performance of the network can be significantly increased. Moreover, the proposed algorithm adjusts throughput and delay based on the assigned weight factor rather than an absolute distinction between high-priority packets and low-priority packets.

The contribution of this work is summarized as follows:

- A novel distributed scheduling scheme for an ad-hoc network is proposed, where both the load-balancing among neighboring nodes and the preferential processing for high importance packets are considered.
- An intra-node slot reallocation algorithm is proposed. Each node is equipped with multiple queues, and this algorithm re-arranges the slot allocation between the queues inside a node. Moreover, this algorithm enables a flexible adjustment of throughput and delay, reflecting assigned weight factors.
- Self-fairness for packets with unequal importance is introduced. This metric incorporates both the weight factor and traffic load. The metric plays an important role in achieving a fairness among the packets with the same weight factor and in supporting service differentiation among packets with different weight factors. It is validated that the proposed scheduling scheme substantially increases the performance of the network.
- It is confirmed that the proposed node scheduling outperforms the absolute priority-based scheduling scheme in terms of delay and throughput. This result is supported by thorough simulation studies accommodating various operation scenarios.

The remainder of this paper is organized as follows: Section 2 describes the various distributed resource allocation medium access control (MAC) protocols proposed in the literature. Section 3 describes the proposed algorithm. In Section 4, the performance of the proposed algorithm is analyzed based on an extensive simulation study, and, finally, Section 5 presents some observational conclusions.

2. Related Works

In [22], the authors proposed a distributed randomized (DRAND) time division multiple access (TDMA) scheduling algorithm, which is a distributed version of the randomized (RAND) time slot scheduling algorithm [23]. DRAND operates in a round-by-round manner and it does not require time synchronizations on the round boundaries, resulting in energy consumption reduction. In this scheme, there are four states for each node: IDLE, REQUEST, GRANT, and RELEASE. Each node is assigned a slot that does not cause a collision within the 2-hop neighboring nodes by sending a state message to the neighboring nodes. The basic idea of the deterministic distributed TDMA (DD-TDMA) [24] is that each node collects information from its neighboring nodes to determine slot allocations. DD-TDMA is superior to DRAND in terms of running time and message complexity. This feature increases energy efficiency because DD-TDMA does not need to wait for a GRANT message, which is transmitted as a response of REQUEST message and it contains a slot allocation permission for unused slots. However, DRAND and DD-TDMA do not consider load balancing and fairness among the nodes.

Algorithms for allocating resources based on the states of networks and nodes were proposed in [25–28]. In [25], a load balancing algorithm for TDMA-based node scheduling was proposed. This scheme makes the traffic load semi-equal and improves fairness in terms of delay. In adaptive topology and load-aware scheduling (ATLAS) [26], nodes determine the amount of resources to be allocated through resource allocation (REACT) algorithms, where each node auctions and bids on time slots. Each node acts as both an auctioneer and a bidder at the same time. During each auction, an auctioneer updates an offer (maximum available capacity) and a bidder updates a claim (capacity to bid in an auction). Through this procedure, resources are allocated to the nodes in a maximum-minimum manner [17]. In [27], an algorithm consisting of two sub-algorithms was proposed. The first is a fair flow vector scheduling algorithm (FFVSA) aiming to improve fairness and optimize slot allocation by considering the active flow requirements of a network. FFVSA uses a greedy collision vector method that has less complexity than the genetic algorithm. The second is a load balanced fair flow vector scheduling algorithm (LB-FFVSA), which increases the fairness of the amount of allocated resources among nodes. In [28], the fairness among nodes was improved in terms of energy consumption through an upgraded version of DRAND. Energy-Topology (E-T) factor was adopted as a criterion for allocating time slots, and E-T-DRAND algorithm was proposed to request time slots. Instead of the randomized approach of DRAND, E-T-DRAND algorithm provides high priority to the nodes with high energy consumption and low residual energy due to the large number of neighboring nodes. E-T-DRAND balances the energy consumption among nodes and enhances scheduling efficiency. Each node determines the number of slots to be reallocated using the number of packets accumulated in the queue of its 1-hop neighboring nodes and the number of allocated slots for these nodes. The slot reallocation procedure must check whether a slot is shared by nodes within 2-hop distance. As a result, the load between nodes becomes semi-equal, and the nodal delay is reduced.

In [29–33], scheduling schemes considering priority were proposed. In [29], for the purpose of reducing delay of emergency data, energy and load balanced priority queue algorithm (ELBPQA) was proposed. In this scheme, four different priority levels are defined according to the position of a node in a network. In [30], the highest priority is given to real-time traffic, and the other priority levels are given to non-real time traffics. In order to reduce the end-to-end delay, the packets with the highest priority are processed in a preemptive manner. In [31], priority- and activity-based QoS MAC (PAQMAC) was proposed. In this scheme, the active time of traffic is dynamically allocated according to priority. Specifically, by adopting a distributed channel access scheme, the packet with high priority have reduced back-off and wait times. In [32], I-MAC protocol, which combines carrier sense multiple access (CSMA) and TDMA schemes, was proposed to increase the slot allocation for nodes with high priority. I-MAC consists of a set-up phase and a transmission phase. The set-up phase consists of neighbor discovery, TDMA time-slot allocation using a distributed neighborhood information-based (DNIB) algorithm, local framing for reuse of time slots, and global synchronization for transmission. Nodes with high priority reduce back-off time to increase the opportunity of winning slot allocation, and nodes with the same priority compete for slot allocation. This scheme reduces the energy consumption of nodes with high priority.

In [33], a QoS-aware media access control (Q-MAC) protocol composed of both intra-node and inter-node scheduling was proposed. Intra-node scheduling determines the priority of packets arriving at the queue of a node. Priority is determined according to the importance of packets and the number of hops to a destination node. Q-MAC consists of five queues, where a queue called an instant queue transmits packets as soon as they arrive. The remaining queues transmit packets following the maximum-minimum fairness principle. Inter-node scheduling is a scheme of data transmission among nodes sharing the same channel. A power conservation MACAW (PC-MACAW) protocol based on the multiple access with collision avoidance protocol for Wireless LANs (MACAW) is applied to schedule data transmission. Q-MAC guarantees QoS through dynamic priority assignment; however, latency can be increased due to heavy computational complexity [34].

A comparative analysis of the protocols mentioned in this section is summarized in Table 1. It is largely classified into with and without prioritizations. In the load-balancing classification, "High" means the clear load-balancing by adopting max-min fairness criterion; "Medium" is an indirect load-balancing method by adjusting idle time and access time; and "Low" is the case where the load-balancing method and its effects are not clearly addressed. In the weight factor classification, "No" is strict priority without quantitative values, and PAQMAC and Q-MAC assign quantitative weight values to packets.

One of the representative fairness measurement methods is Jain's fairness index, which is a value range (0, 1), and the closer it is to 1 the fairer it is [20]. Jain's fairness index can measure the fairness of an entire system in a relatively simple way, but it cannot measure the fairness of nodes to which a weight factor is assigned. In [21], the authors proposed a quantitative fairness measurement method applicable to scheduling algorithms with unequal weight factors.

Table 1. Comparative analysis of related works.

Classification	Protocol	Access Mechanism	Load-Balancing	Weight Factor	Goal
Without Prioritization	DRAND [22]	TDMA	No	N/A	To allocate resources efficiently in ad-hoc networks
	LocalVoting [25]	TDMA	High	N/A	To decrease average delay by making the load between neighbor nodes semi-equal
	ATLAS [26]	TDMA	High	N/A	To adapt topology changes fast and allocate resources considering neighbor nodes
With Prioritization	ELBPQA [29]	CSMA/CA	Low	No	To minimize delay of high priority packets
	Algo. [30]	TDMA	Low	No	To minimize end-to-end delay of high priority packets
	I-MAC [32]	CSMA + TDMA	Medium	No	To increase chance of resource allocation for high priority nodes by CSMA + TDMA
	PAQMAC [31]	CSMA/CA	No	Quantitative	To allocate active time dynamically by considering the priority of packets
	Q-MAC [33]	CSMA/CA	Medium	Quantitative	To increase energy efficiency while providing service differentiation

3. Proposed Node Scheduling with Weight Factor

Instead of conventional absolute priority-based scheduling, an adjustable and flexible scheduling scheme is proposed. This scheme reallocates slots by taking the weights assigned to the queues of nodes into account. Specifically, intra-node scheduling, which reallocates slots between the queues for high- and low-importance packets, is introduced. Then, it is followed by inter-node scheduling adopted from [25], which reallocates slots among neighboring nodes to increase the fairness measured in terms of traffic load.

The proposed algorithm consists of three steps: (1) free time slot allocation, which is a process of allocating the initialized slots (unallocated empty slots) to packets; (2) the intra-node slot reallocation algorithm, which exchanges slots between the queues of a node with different importance values using self-fairness; and (3) the inter-node slot reallocation among 1-hop neighbors using a load balancing algorithm (slot exchange between queues with the same importance). The procedure of this algorithm is depicted in Figure 1.

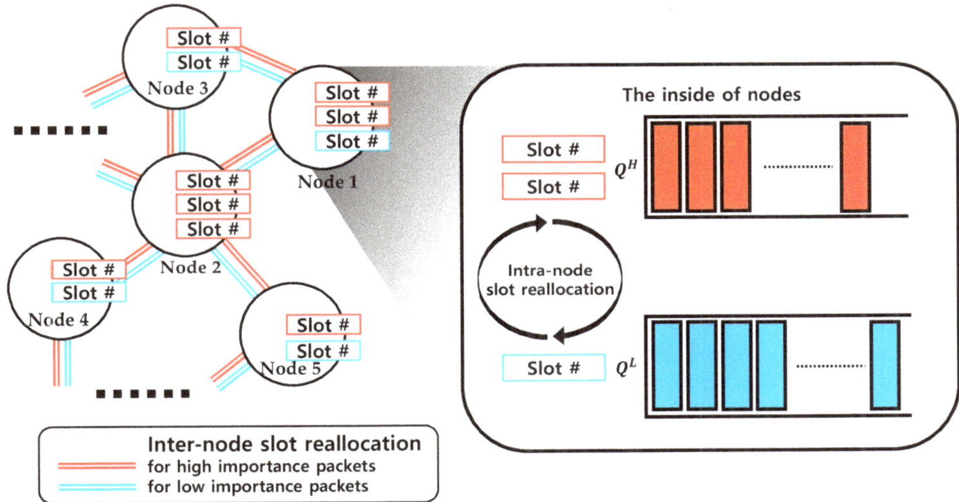

Figure 1. Intra-node slot reallocation and inter-node slot reallocation of the proposed scheduling algorithm.

All the nodes have two types of queues for storing packets of different importance. Q^H and Q^L are queues for high- and low-importance packets, respectively. Q^A, $A \in \{H, L\}$ represent Q^H or Q^L according to the indicator A, respectively. In the following, A is used as an indicator representing importance. The number of slots required to transmit all the packets at Q^A of node i at frame time t is represented by $q_t^{(A,i)}$, and the number of slots assigned to Q^A of node i at frame time t for packet transmission is represented by $p_t^{(A,i)}$. Assuming that the packet and the slot sizes are the same, $q_t^{(A,i)}$ is equal to the number of packets in Q^A. $p_t^{(A,i)}/q_t^{(A,i)}$ is the inverse load of Q^A and expressed as $X_t^{(A,i)} = p_t^{(A,i)}/q_t^{(A,i)}$.

Free time slot allocation requires REQUEST and RELEASE messages exchanges, as in DRAND. The number of packets to be transmitted by node i is $q_t^{(H,i)} + q_t^{(L,i)}$, and node i with $q_t^{(H,i)} + q_t^{(L,i)} > 0$ can be allocated slots that are not reserved by the nodes within 2-hop distance. Note that the nodes within 2-hop distance cannot reuse time slot to avoid packet collisions and this reuse can be prevented by slot reallocation between 1-hop nodes. Node i allocates as many as $q_t^{(H,i)}$ slots to Q^H and increases $p_t^{(H,i)}$ by the number of allocated slots. If $q_t^{(H,i)} = p_t^{(H,i)}$, Q^H does not need to be allocated more slots; accordingly, the slots are allocated to Q^L, and $p_t^{(L,i)}$ is increased. Afterwards, $p_t^{(H,i)}$ and $p_t^{(L,i)}$ are reallocated through the intra-node slot reallocation algorithm. If both Q^H and Q^L are allocated as many as $q_t^{(H,i)}$ and $q_t^{(L,i)}$, no more slots are assigned.

In the intra-node slot reallocation, a self-fairness index is used to reallocate packets between Q^H and Q^L of each node. Self-fairness is a measure of how fairly an amount of "resources" is assigned to a particular node by considering the weight assigned to that node. In this measurement, the resource can be bandwidth, time slots, etc. The proposed algorithm uses inverse load $X_t^{(A,i)}$ as a resource for self-fairness measurement.

In the proposed algorithm, self-fairness applies to two different queues of each node. Hence, each node has two self-fairness values for its two queues (Q^H and Q^L). The self-fairness value for Q^A of node i is denoted by $F_t^{(A,i)}$ and defined as it is presented in Equations (1)–(3) [21]:

$$F_t^{(A,i)} = \frac{\log\left(\varphi_t^{(A,i)}\right)}{\log\left(r^{(A,i)}/r_{Tot}^{(A,i)}\right)}, A \in \{H, L\} \tag{1}$$

$$\varphi_t^{(A,i)} = \frac{X_t^{(A,i)}}{\sum_{k \in \mathcal{N}_i} X_t^{(H,k)} + \sum_{k \in \mathcal{N}_i^{(1)}} X_t^{(L,k)}} \tag{2}$$

$$r_{Tot}^{(A,i)} = \sum_{k \in \mathcal{N}_i} r^{(H,k)} + r^{(L,k)} \tag{3}$$

where $\varphi_t^{(A,i)}$ is the ratio of resources allocated to Q^A at node i to the sum of resource allocated to Q^H and Q^L at 1-hop neighboring nodes, $\mathcal{N}_i$ is a set of 1-hop neighbor nodes of node i, $r^{(A,i)}$ is the weight assigned to Q^A of node i, and $r_{Tot}^{(A,i)}$ is the sum of the weights of 1-hop neighboring nodes. When the weight is high, more slots are allocated to increase the inverse-load, resulting in a fairer resource allocation. By setting $r^{(H,i)} > r^{(L,i)}$, more important packets are allocated more slots than less important packets. Accordingly, $F_t^{(A,i)}$ is a quantitative value for Q^A of node i, indicating whether the load of Q^A is high or low considering the weight assigned. Therefore, it is used as an index to compare the fairness of slot allocation with unequal weight factor.

When $F_t^{(A,i)} = 1$, the allocation is in the fairest state. When the amount of slots allocated is small compared to the assigned weight factor, $F_t^{(A,i)} > 1$ can be satisfied because $\varphi_t^{(A,i)} \in [0, 1]$. In this case, it is necessary to gain more slots from the other queue. In the opposite case, if too many slots are allocated, $F_t^{(A,i)} < 1$ can be satisfied, and Q^A must release its own slots. When a slot is gained, $p_t^{(A,i)}$ and $\varphi_t^{(A,i)}$ will increase, resulting in a decrement of $F_t^{(A,i)}$. In contrast, when a slot is released, $F_t^{(A,i)}$ increases. The intra-node slots reallocation algorithm adjusts $F_t^{(H,i)}$ and $F_t^{(L,i)}$ to be as close to 1 as possible, which improves the self-fairness. Specifically, when $F_t^{(H,i)} > F_t^{(L,i)}$, the slots allocated to Q^L are released to Q^H, and vice versa when $F_t^{(H,i)} < F_t^{(L,i)}$. The algorithm for $F_t^{(H,i)} > F_t^{(L,i)}$ is detailed in Algorithm 1.

In Algorithm 1, $\hat{F}_t^{(H,i)}$ and $\hat{F}_t^{(L,i)}$ are the expected self-fairness values calculated assuming that slots are reallocated. It is assume that Q^H gains a slot from Q^L, hence, $\hat{F}_t^{(H,i)}$ is calculated by increasing $p_t^{(H,i)}$ by 1, and $\hat{F}_t^{(L,i)}$ is calculated by decreasing $p_t^{(L,i)}$ by 1. The updated $p_t^{(H,i)}$ and $p_t^{(L,i)}$ are transmitted to its 1-hop neighboring nodes at the end of each frame. Accordingly, during slot exchange at frame time t, φ is calculated using only the locally updated $p_t^{(H,i)}$ and $p_t^{(L,i)}$ by intra-node slot exchange. In the next frame, the self-fairness is updated through information exchanges among neighboring nodes. When $p_t^{(L,i)} = 1$ and Q^L releases a slot, $p_t^{(L,i)}$ will be 0. This makes $\varphi_t^{(L,i)} = 0$, and $\hat{F}_t^{(L,i)}$ becomes infinity. To prevent this, a minimum default value above 0 is assigned to $p_t^{(L,i)}$ under this situation.

Algorithm 1. Increasing Q^H slot allocation

1: **for** all node i **do**
2:   **if** $q_t^{(H,i)}$!= 0
3:     Calculate $F_t^{(H,i)}$
4:   **end if**
5:   **if** $q_t^{(L,i)}$!= 0
6:     Calculate $F_t^{(L,i)}$
7:   **end if**
8:   **if** $F_t^{(H,i)} > F_t^{(L,i)}$
9:     **if** $p_t^{(L,i)} > 0$
10:       $\hat{F}_t^{(H,i)} \leftarrow p_t^{(H,i)} + 1$
11:       $\hat{F}_t^{(L,i)} \leftarrow p_t^{(L,i)} - 1$
12:       $\hat{\mathcal{F}}_t^i \leftarrow \sqrt{\left(1 - \hat{F}_t^{(H,i)}\right)^2 + \left(1 - \hat{F}_t^{(L,i)}\right)^2}$
13:       $\mathcal{F}_t^i \leftarrow \sqrt{\left(1 - F_t^{(H,i)}\right)^2 + \left(1 - F_t^{(L,i)}\right)^2}$
14:     **end if**
15:     **while** $\mathcal{F}_t^i > \hat{\mathcal{F}}_t^i$ **do**
16:       $p_t^{(H,i)} \leftarrow p_t^{(H,i)} + 1$
17:       $p_t^{(L,i)} \leftarrow p_t^{(L,i)} - 1$
18:       **if** $p_t^{(L,i)} > 0$
19:         $\mathcal{F}_t^i \leftarrow \hat{\mathcal{F}}_t^i$
20:         $\hat{\mathcal{F}}_t^i \leftarrow \sqrt{\left(1 - \hat{F}_t^{(H,i)}\right)^2 + \left(1 - \hat{F}_t^{(L,i)}\right)^2}$
21:       **else break;**
22:       **end if**
23:     **end while**
24:   **end if**
25: **end if**

At every frame, slots are reallocated until self-fairness can no longer be improved. Note that the fairness index 1 is the fairest state. Consequently, the Euclidean distance between the fairest status $F_t^{(H,i)} = F_t^{(L,i)} = 1$ and a current $(F_t^{(H,i)}, F_t^{(L,i)})$ combination is introduced as a metric representing a target fairness, as it is presented in Equation (4):

$$\mathcal{F}_t^i = \sqrt{\left(1 - F_t^{(H,i)}\right)^2 + \left(1 - F_t^{(L,i)}\right)^2} \tag{4}$$

Now, the expected Euclidean distance $\hat{\mathcal{F}}_t^i$ from the expected fairness $(\hat{F}_t^{(H,i)}, \hat{F}_t^{(L,i)})$ is compared with the current Euclidean distance $\mathcal{F}_t^i$ from $(F_t^{(H,i)}, F_t^{(L,i)})$. If $\hat{\mathcal{F}}_t^i < \mathcal{F}_t^i$, Q^H gains a slot from Q^L, and $p_t^{(H,i)}$ and $p_t^{(L,i)}$ are updated. Because slot reallocation is an intra-node process, collisions with 2-hop neighboring nodes need not be considered.

When $F_t^{(H,i)} < F_t^{(L,i)}$, the slot reallocation algorithm is very similar to Algorithm 1, and $\hat{F}_t^{(H,i)}$ and $\hat{F}_t^{(L,i)}$ are calculated with $p_t^{(H,i)} - 1$ and $p_t^{(L,i)} + 1$, respectively. However, instead of $p_t^{(L,i)} > 0$ in lines 9 and 18 of Algorithm 1, $p_t^{(H,i)} > 1$ is used as a slot release condition. This prevents $p_t^{(H,i)}$ from being zero by releasing all slots to Q^L to improve the fairness when $q_t^{(H,i)} \ll q_t^{(L,i)}$. That is, $p_t^{(H,i)} \geq 1$ is guaranteed in any situation.

After the intra-node slot reallocation algorithm, the inter-node slots reallocation [25] follows. At this time, the slot exchange does not consider the weights of Q^H and Q^L any more because these exchanges take place among the queues with the same importance. Node i's Q^A computes

$u_t^{(A,i)}$ to determine how many slots to reallocate with a 1-hop neighboring node as it is presented in Equation (5) [25]:

$$u_t^{(A,i)} = \left[q_t^{(A,k)} \cdot \frac{\sum_{k \in \mathcal{N}_i} p_t^{(A,k)}}{\sum_{k \in \mathcal{N}_i} q_t^{(A,k)}} \right] - p_t^{(A,k)} \tag{5}$$

If $u_t^{(A,i)} > 0$, slots are gained from the 1-hop neighboring node. If $u_t^{(A,i)} < 0$, slots are released to the 1-hop neighboring node. The number of reallocated slots is determined by $\min\{u_t^{(A,i)}, u_t^{(A,i)} - u_t^{(A,k)}, p_t^{(A,k)}\}$. This increases the equality of the inverse-load of the same importance among node i and its 1-hop neighboring nodes. These processes are performed for all nodes in a node-by-node manner. The same intra-node and inter-node slot reallocations are repeated in the next frame.

4. Performance Evaluation

A network simulator [35] implemented in Java was used for performance analysis of the proposed algorithm. No isolated nodes are assumed, i.e., all the nodes have at least a single 1-hop neighbor node. Accordingly, in establishing a connection, any two nodes can be connected with each other through multi-hop links. The connections are established using arbitrarily chosen pairs of a source node and a destination node, and high- and low-importance connections generate high- and low-importance packets, respectively. In the following, high- and low-importance packets are denoted by Pkt^H and Pkt^L, respectively.

For the performance analysis, the throughput, delay, and fairness are measured by changing the connection creation ratio (between Pkt^H and Pkt^L) and the weight factor setting. Then, the proposed algorithm is compared with the absolute priority-based algorithm in which Pkt^H preempts time slots when allocating free time slots. Note that the absolute priority algorithm adopts only the inter-node slot reallocation algorithm, not the intra-node slot one.

The generation ratios of high- and low-importance connections are denoted by α, $1 - \alpha \in [0,1]$. The weight factor setting in Q^A is denoted by r^A. Assuming that Q^H and Q^L of all nodes have the same weight settings as r^H and r^L, respectively, the node index i can be dropped from the weight factors. The weight factors are set as: r^H, $r^L \in [0,10]$ and $r^H + r^L = 10$.

The performance of the proposed scheme was measured in two scenarios. Table 2 lists the parameters setting for each scenario. In the first scenario, a fixed number of connections are created at the starting epoch of the simulation, the packets of the connections are generated at fixed time intervals, and the number of packets generated for each connection is the same. In the second scenario, connections are created based on Poisson processes. Unlike the first scenario, the number of packets generated per connection follows a Poisson distribution. The arrival rate λ determines the connection creation interval. The duration of each connection follows an exponential distribution of parameter μ, which determines the number of packets generated in each connection. The packets are generated at a fixed interval, as in the first scenario. Each connection is closed if all the packets arrive at its destination node. Because the connections are continually generated, in the second scenario, the simulation duration is specified at the beginning of the simulation. For both scenarios, the final measurement is the average over 1000 independent simulations.

Table 2. Simulation parameters.

Parameter	Value
Number of nodes	30
Transmission range	5 units
Topology size	50×50 units
Frame length	30 time-units
Packet generation interval	5 time-units
Number of connections N_c	50–500
Number of packets per connection N_p	50
Connection duration μ	10^3 time-units
Arrival rate λ	10^{-3}–10^{-1} time-units^{-1}
Simulation time T (first/second scenario)	3000/150,000 time-units

In the first scenario, the performance of the proposed algorithm was analyzed with the increasing total number of connections and the various settings of the weight factor and α. The total number of created connections is the sum of the high- and low-importance connections. Throughput, packet delivery ratio, 1-hop delay, and fairness are measured and compared with those of absolute priority-based scheduling. Throughput refers to the number of all packets arriving at a destination node during the simulation. However, in the first scenario, since the number of generated connections is determined at the beginning of the simulation, the throughput measured when all packets arrive at a destination node will be simply the product of N_c (number of connections) and N_p (number of generated packets per connection). Therefore, throughput is measured not at the end of the simulation but at a predefined time T, which is large enough for the transmission of packets in the network to be in a steady state. The packet delivery ratio means the proportion of received packets to the packets sent. The 1-hop delay is measured as the average of ((the time when a packet is dequeued) minus (the time when a packet is enqueued)). The results of the absolute priority-based algorithm are marked as Preempt.Pkt^H and Preempt.Pkt^L.

Figures 2–6 show the results of the first scenario. Figure 2 depicts the throughputs with the increasing total number of connections, various weight factors, and $\alpha = 0.3$. When the number of connections is small, most packets are delivered to the destination nodes until the predefined time T because the network is not heavily loaded. For this reason, in Figure 2a,b, when the number of connections is 50, the throughput of Pkt^H is lower than that of Pkt^L because the number of Pkt^H is lower than Pkt^L. In most cases, if the number of connections increases, the throughput of Pkt^H is higher than that of Pkt^L. However, in Figure 2b, when the weight factors are $r^H = 7$ and $r^L = 3$, the throughput of Pkt^L is higher than that of Pkt^H, even when the number of connections increases. Note that the proposed algorithm considers not only the weight factors but the traffic load as well; hence, even when $r^L < r^H$, the throughput of Pkt^L is higher than that of Pkt^H in the entire range of N_c. The service differentiation between Pkt^H and Pkt^L is more evidently shown in Figure 2c,d. As shown in these figures, over all the range of the number of connections, the packet delivery ratio of Pkt^H is higher than Pkt^L. Specifically, Figure 2b with $r^H = 7$, $r^L = 3$ can be compared with Figure 2d with $r^H = 7$, $r^L = 3$. In this case, Figure 2b shows that the throughput of Pkt^L is higher than Pkt^H. However, Figure 2d shows that the packet delivery ratio of Pkt^H is still twice as high as that of Pkt^L. This result clearly shows that the proposed scheme preferentially processes packets reflecting the weight factors. When the absolute priority-based algorithm is applied, as the number of Pkt^H to be transmitted increases owing to the increment of the number of connections, the opportunity for Pkt^L slot allocation decreases, resulting in a further decrease in the throughput of Pkt^L.

In Figure 3, throughputs are measured when $r^H \cdot \alpha = r^L \cdot (1-\alpha)$ is satisfied under the condition of increasing number of connections. Figure 3 shows the characteristics of the proposed algorithm by considering both the weight factor and traffic load. When $r^H \cdot \alpha = r^L \cdot (1-\alpha)$ is satisfied, it is confirmed that the throughputs of Pkt^H and Pkt^L have similar values and converge to a single value, as shown in Figure 3.

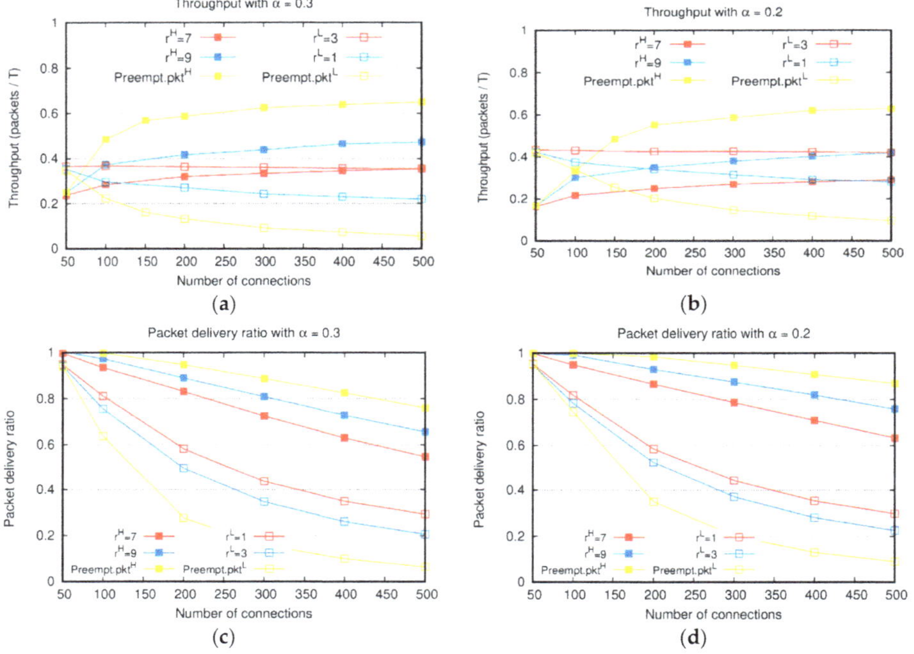

Figure 2. Throughput comparisons between Pkt^H and Pkt^L with increasing number of connections: (**a**,**b**) throughputs with $\alpha = 0.3$, $\alpha = 0.2$; (**c**,**d**) packet delivery ratios with $\alpha = 0.3$, $\alpha = 0.2$.

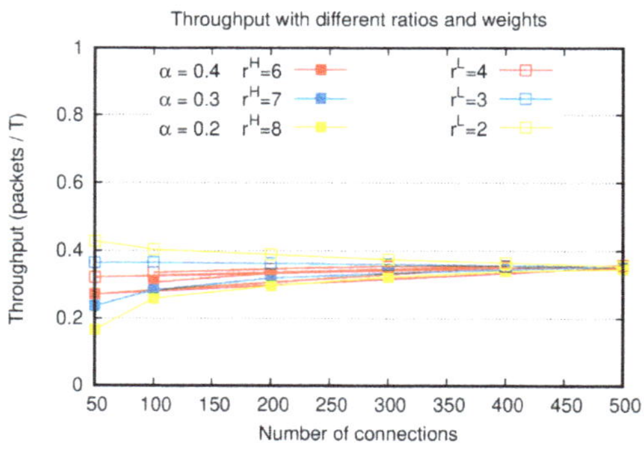

Figure 3. Throughput with various connection creation ratios and weight factors with increasing number of connections.

As shown in Figures 2 and 3, the sums of the throughputs of Pkt^H and Pkt^L are similar when N_c is the same, even though α and the weight factors are different. This is because even when the number of allocated slots of Pkt^H and Pkt^L are changed by α and the weight factors during the process of reallocation, the number of allocated slots in the entire network does not change. Therefore, there is a tradeoff between the throughputs of Pkt^H and Pkt^L depending on the weight factors. From Figures 2 and 3, it is confirmed that an appropriate weight factor setting is necessary to adjust the throughputs of Pkt^H and Pkt^L for various network situations with different α.

Figure 4 shows 1-hop delay with various weight factors and α with the increasing total number of connections. Similar to in Figures 2 and 3, when the number of connections is small, all the generated packets can be delivered to destination nodes, resulting in nearly no difference in the delay between Pkt^H and Pkt^L. However, as the number of connections increases, the delays of both Pkt^H and Pkt^L increase, and the delay difference between Pkt^H and Pkt^L becomes conspicuous. Compared to the absolute priority-based algorithm, the delay gap between Pkt^H and Pkt^L of the proposed algorithm is relatively small. In the case of $r^H = 7$ and $r^L = 3$ shown in Figure 4a, when N_c is 500, the delay of Pkt^L is twice that of Pkt^H. On the other hand, the delay of Preempt. Pkt^L is more than 6 times the delay of Preempt.Pkt^H. The delay of Pkt^H increases compared to Preempt.Pkt^H, but the delay of Pkt^L decreases much more than Preempt.Pkt^L. In particular, when $r^H = 9, r^L = 1$, and $N_c = 500$ in Figure 4b, the delay of Pkt^H increases by approximately 500 time slots compared to Preempt.Pkt^H, but the delay of Pkt^L decreases by approximately 3000 time slots compared to Preempt.Pkt^L, and it is a noticeable improvement. The average sum delay of Pkt^H and Pkt^L is reduced by 20% compared to that of Preempt.Pkt^H and Preempt.Pkt^L. This means that, compared to the absolute priority-based algorithm, the proposed algorithm achieves the higher performance. Moreover, the proposed algorithm can achieve the same delay performance with Preempt.Pkt^H by throttling Pkt^L, i.e., with $r^H = 10$ and $r^L = 0$. When $\alpha = 0.5$, the number of Pkt^H to be transmitted increases and the delay of Pkt^H, at the same N_c, increases compared to the case of $\alpha = 0.3$. In the whole range of N_c, the delay of Pkt^H in Figure 4b is higher than that of Pkt^H in Figure 4a. In addition, Pkt^H's delay when $r^H = 7$ in Figure 4a and that when $r^H = 9$ in Figure 4b are similar.

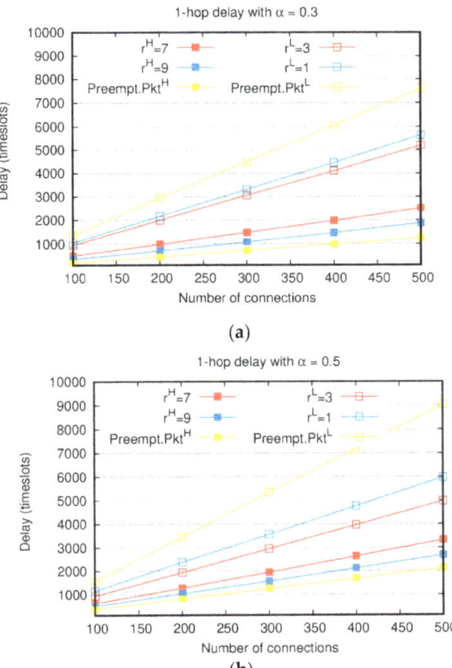

Figure 4. Delay comparison between priority-based algorithms Pkt^H and Pkt^L: (**a**) 1-hop delay with $\alpha = 0.3$; (**b**) 1-hop delay with $\alpha = 0.5$.

In Figures 2 and 4, for Pkt^H, the higher r^H is, the better the performances of throughput and delay are. The decrement in r^L due to the increased r^H leads to the worse performance of throughput and delay of Pkt^L. The larger the difference between the values of r^H and r^L, the larger the performance gap between the throughput and delay of Pkt^H and Pkt^L. This confirms that Pkt^H and Pkt^L are flexibly adjusted based on the values of the weight factor in various network situations.

In Figure 5, the proposed scheduling scheme is compared with DRAND, LocalVoting, and Q-MAC. Q-MAC was developed for CSMA/CA and the packets with high weight value had a relatively high probability of accessing channel. For comparison, Q-MAC was modified to be applicable to TDMA. Specifically, the slots of Q-MAC are initialized according to the weight values, and the inter-node reallocation of LocalVoting is followed. As shown in Figure 5a, the delay of Pkt^H is better than both DRAND and LocalVoting, and slightly worse than Q-MAC with Pkt^H. Even Pkt^L shows the better performance than DRAND and slightly worse than LocalVoting. Specifically, the delay of DRAND is twice longer than Pkt^L and four times longer than Pkt^H. LocalVoting shows the performance better than DRAND through the neighbor-aware load balancing. However, the proposed scheme of Pkt^H still outperforms LocalVoting. The delay of Pkt^H is 1.8 times smaller than LocalVoting. In Figure 5b, the average delay of the proposed scheme shows the best performance. Q-MAC and LocalVoting show the similar performance with each other. In Figure 5c, the throughput of the proposed scheme with Pkt^H lower than Q-MAC with Pkt^H. However, the throughput of the proposed scheme with Pkt^L is higher than Q-MAC with Pkt^L. Note that the throughput of LocalVoting in Figure 5c is the sum of its Pkt^H and Pkt^L. In Figure 5d, the proposed scheme achieves the highest throughput. In Figure 5b,d, it is ensured that the proposed scheme possesses the excellent performance in slot allocation because it achieves the highest throughput and the lowest delay.

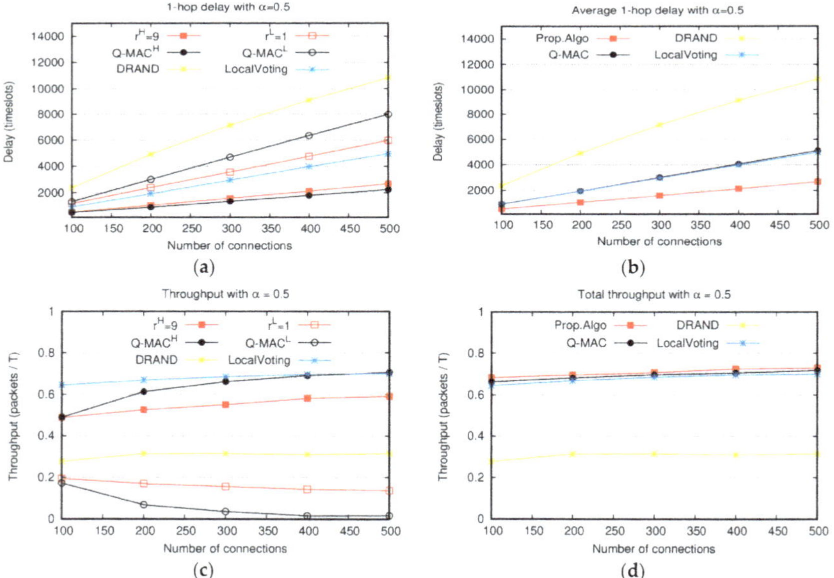

Figure 5. Delay and throughput comparison between the proposed algorithm and other scheduling algorithms with $\alpha = 0.5$: (**a**,**b**) 1-hop delays of different weight values and average 1-hop delay; (**c**,**d**) throughputs with different weight values and total throughputs.

Moreover, Figure 5a,c show that the service differentiation of the proposed scheme is enabled compared with other schemes. These are the major contributions of the proposed scheme.

Figure 6 compares Jain's fairness [20] of Pkt^H and Pkt^L with and without the proposed algorithm. In this figure, in terms of $\Gamma^{(A,i)}$, Jain's fairness index shows how fairly resources are allocated among the queues of the same importance. $\Gamma^{(A,i)}$ is the ratio of the accumulative number of packets transmitted from a queue to the number of accumulated packets in a queue until T, which can be expressed as Equation (6). Similar to the throughput measurement, at the end of the simulation, all packet delivery is completed; accordingly, Jain's fairness is calculated at time T.

$$\Gamma^{(A,i)} = \frac{\sum_{t=0}^{T} p_t^{(A,i)}}{\sum_{t=0}^{T} q_t^{(A,i)}}, \ A \in \{H, L\} \tag{6}$$

In this analysis, $\alpha = 0.3$ and $r^H = 7$, $r^L = 3$ are considered. When the number of connections is small, the fairness index is high regardless of the adoption of the proposed algorithm because the $\Gamma^{(A,i)}$ of most nodes becomes close to 1. For the absolute priority-based algorithm, as the number of connections increases, only a few nodes are allocated slots for Preempt.Pkt^L. Since most nodes cannot transmit Preempt.Pkt^L, the fairness of Preempt.Pkt^L is very low. In contrast, when the intra-node slot reallocation of the proposed algorithm is adopted, time slots proportional to r^L are allocated to Q^-, and this results in an increase in the fairness index. As a result, the fairness performance of Pkt^L is significantly increased compared to that of Pkt^H when the intra-node slot exchange algorithm is applied.

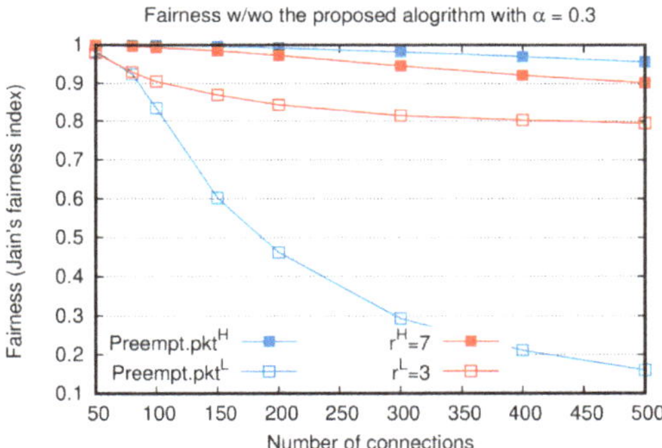

Figure 6. Jain's fairness comparison between the proposed algorithm and absolute priority-based scheduling.

Figure 7 shows the delay and throughput performance of the second scenario with the increasing Poisson arrival rate λ. In Figure 7a,b, because $\alpha = 0.5$ is applied, the numbers of Pkt^H and Pkt^L are similar. Although the connection creation interval and the number of packets generated for each connection are varied, Figure 7 shows similar performances to those of the first scenario. The larger the difference between r^H and r^L, the greater the performance gap between Pkt^H and Pkt^L. For instance, in Figure 7a, when the arrival rate is 0.01 time $-$ units^{-1} and the weight factors are $r^H = 7$ and $r^L = 3$, the Pkt^L delay is approximately 1.5 times longer than the Pkt^H delay. However, when the weight factors are $r^H = 9$ and $r^L = 1$, Pkt^L delay is over two times Pkt^H delay. When the arrival rate is low, the connection creation interval is long, and the number of connections created during the entire simulation is small. As shown in Figure 7a,b, when the arrival rates are as low as 0.001 and 0.002 time $-$ units^{-1}, there is only a slight difference in delay and throughput between Pkt^H and Pkt^L regardless of the weight factor setting.

Figure 7c shows the throughput when the number of Pkt^L is larger than that of Pkt^H, by setting $\alpha = 0.3$. The result of Figure 7c is very similar to that of Figure 2a when N_c ranges between 100 and 500. In particular, if $r^H \cdot \alpha = r^L \cdot (1 - \alpha)$ is satisfied by setting $r^H = 7$, $r^L = 3$, the throughputs of Pkt^H and Pkt^L converge to a constant value. However, note that α is set as 0.3, i.e., 70% of the generated packets are Pkt^L and the remaining 30% is Pkt^H. Even in this asymmetric packet generation scenario, Pkt^H achieves the higher throughput than Pkt^L. Accordingly, this clearly shows that the service differentiation between Pkt^H and Pkt^L is attained.

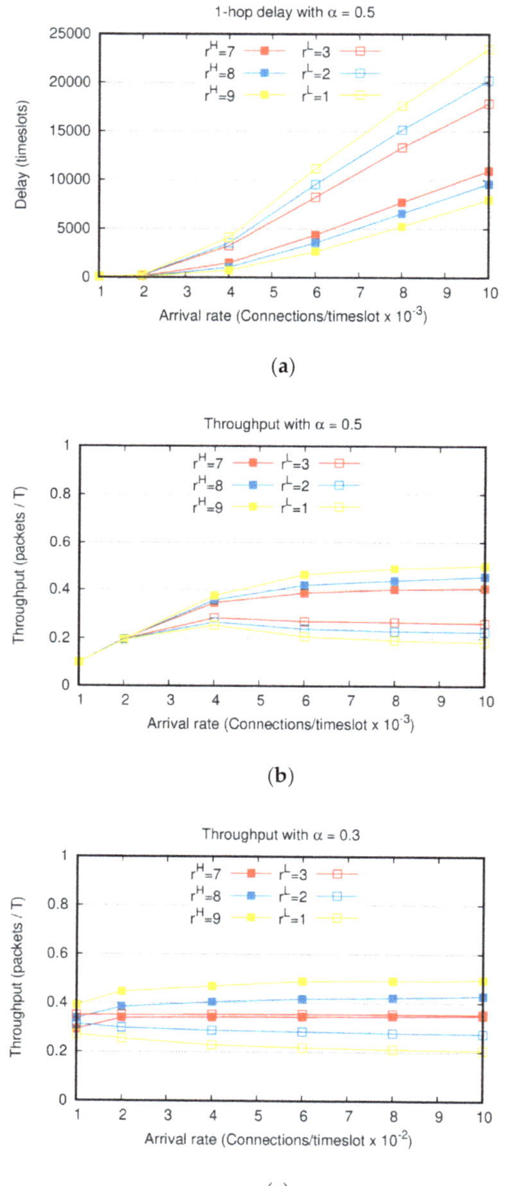

(a)

(b)

(c)

Figure 7. Delay and throughput with increasing Poisson arrival rates and the same weight factor setting: (**a**) 1-hop delay with $\alpha = 0.5$; (**b**) throughput with $\alpha = 0.5$; (**c**) throughput with $\alpha = 0.3$.

5. Conclusions

In this paper, a novel distributed node scheduling algorithm for an ad-hoc network was proposed. This scheme flexibly adjusts time slot allocations according to weight factor and traffic load. From thorough simulation studies under various environments, the performance differentiation reflecting weight factor setting was validated. It was confirmed that, as the weight of the high importance packets increases, the delay decreases and the throughput at the same time increases. Because the proposed algorithm considers both the weight factors and traffic loads, even the throughput and delay for the same weight factors can be adjusted separately according to the connection creation ratios with different importance. Through comparison with other distributed node scheduling algorithms, the advantages of the proposed algorithm were validated. Specifically, it supports load balancing with neighboring nodes and preferential processing of important data. In addition, compared to the conventional absolute priority-based algorithm, the proposed algorithm shows performance improvement in terms of throughput, delay, and fairness for low-importance packets. Moreover, the performance comparison with other scheduling scheme ensures the excellent performance of the proposed scheme because it achieves the highest throughput and the lowest delay. These results verify that both the service differentiation and performance improvement can be achieved through an appropriate weight factor setting.

Author Contributions: W.L. and T.K. (Taejoon Kim) conceived and designed the experiments; W.L. and T.K. (Taehong Kim) performed the network simulation; W.L, T.K. (Taehong Kim), and T.K. (Taejoon Kim) analyzed the data; T.K. (Taejoon Kim) acquired funding; W.L. and T.K. (Taejoon Kim) wrote the paper. All authors have read and agreed to the published version of the manuscript.

Funding: This research was supported by Basic Science Research Program through the National Research Foundation of Korea (NRF) funded by the Ministry of Education (2020R1I1A3068305).

Conflicts of Interest: The authors declare no conflict of interest.

References

1. Rajaraman, R. Topology control and routing in ad hoc networks. *SIGACT News* **2002**, *33*, 60. [CrossRef]
2. Sharmila, S.; Shanthi, T. A survey on wireless ad hoc network: Issues and implementation. In Proceedings of the 2016 International Conference on Emerging Trends in Engineering, Technology and Science (ICETETS), Pudukkottai, India, 24–26 February 2016; pp. 1–6.
3. Wu, K.; Harms, J. Multipath routing for mobile ad hoc networks. *J. Commun. Netw.* **2002**, *4*, 48–58. [CrossRef]
4. Fossa, C.E.; Macdonald, T.G. Internetworking tactical MANETs. In Proceedings of the 2010-MILCOM 2010 Military Communications Conference, San jose, CA, USA, 31 October–3 November 2010; pp. 611–616.
5. Bekmezci, I.; Alagöz, F. A New TDMA Based Sensor Network for Military Monitoring (MIL-MON). In Proceedings of the MILCOM 2005 IEEE Military Communications Conference, Atlantic City, NJ, USA, 17–20 October 2006; pp. 1–6.
6. Alemdar, H.; Ersoy, C. Wireless sensor networks for healthcare: A survey. *Comput. Netw.* **2010**, *54*, 2688–2710. [CrossRef]
7. Zhang, C.; Zhang, M.; Su, Y.; Wang, W. Smart home design based on ZigBee wireless sensor network. In Proceedings of the 7th International Conference on Communications and Networking in China, Kunming, China, 8–10 August 2012; pp. 463–466.
8. Saleh, N.; Kassem, A.; Haidar, A.M. Energy-Efficient Architecture for Wireless Sensor Networks in Healthcare Applications. *IEEE Access* **2018**, *6*, 6478–6486. [CrossRef]
9. Chao, H.-L.; Kuo, J.-C.; Liao, W. Fair scheduling with QoS support in ad hoc networks. In Proceedings of the 27th Annual IEEE Conference on Local Computer Networks 2002 Proceedings LCN 2002 LCN-02, Tampa, FL, USA, 6–8 November 2003; pp. 502–507.
10. Kim, H.; Min, S.-G. Priority-based QoS MAC protocol for wireless sensor networks. In Proceedings of the 2010 IEEE International Symposium on Parallel & Distributed Processing (IPDPS), Rome, Italy, 23–29 May 2009; pp. 1–8.

11. Ye, W.; Heidemann, J.; Estrin, D. An energy-efficient MAC protocol for wireless sensor networks. In Proceedings of the Twenty-First Annual Joint Conference of the IEEE Computer and Communications Societies, New York, NY, USA, 23–27 June 2003; Volume 3, pp. 1567–1576.

12. Carlos-Mancilla, M.; Fapojuwo, A.; Lopez-Mellado, E.; Siller, M. An efficient reconfigurable ad-hoc algorithm for multi-sink wireless sensor networks. *Int. J. Distrib. Sens. Netw.* **2017**, *13*, 1550147717733390. [CrossRef]

13. Abu Salem, A.O.; Shudifat, N. Enhanced LEACH protocol for increasing a lifetime of WSNs. *Pers. Ubiquitous Comput.* **2019**, *23*, 901–907. [CrossRef]

14. Sakthy, S.S.; Bose, S. Dynamic Model Node Scheduling Algorithm Along with OBSP Technique to Schedule the Node in the Sensitive Cluster Region in the WSN. *Wirel. Pers. Commun.* **2020**, *114*, 1–15. [CrossRef]

15. Zareei, M.; Vargas-Rosales, C.; Hernndez, R.V.; Azpilicueta, E. Efficient Transmission Power Control for Energy-harvesting Cognitive Radio Sensor Network. In Proceedings of the 2019 IEEE 30th International Symposium on Personal, Indoor and Mobile Radio Communications (PIMRC Workshops), Istanbul, Turkey, 8 September 2019; pp. 1–5.

16. Zareei, M.; Vargas-Rosales, C.; Anisi, M.H.; Musavian, L.; Villalpando-Hernandez, R.; Goudarzi, S.; Mohamed, E.M. Enhancing the Performance of Energy Harvesting Sensor Networks for Environmental Monitoring Applications. *Energies* **2019**, *12*, 2794. [CrossRef]

17. Radunovic, B.; Le Boudec, J.-Y. A Unified Framework for Max-Min and Min-Max Fairness with Applications. *IEEE/ACM Trans. Netw.* **2007**, *15*, 1073–1083. [CrossRef]

18. Shi, H.; Prasad, R.V.; Onur, E.; Niemegeers, I.G.M.M. Fairness in Wireless Networks:Issues, Measures and Challenges. *IEEE Commun. Surv. Tutor.* **2013**, *16*, 5–24. [CrossRef]

19. Kelly, F. Charging and rate control for elastic traffic. *Eur. Trans. Telecommun.* **1997**, *8*, 33–37. [CrossRef]

20. Jain, R.; Chiu, D.-M.; Hawe, W. *A Quantitative Measure of Fairness and Discrimination for Resource Allocation in Shared Computer System*; Digit. Equip. Corp. Tech. Rep-301; Eastern Research Laboratory, Digital Equipment Corporation: Hudson, MA, USA, 1984; pp. 1–38.

21. Elliott, R. A measure of fairness of service for scheduling algorithms in multiuser systems. In Proceedings of the IEEE CCECE2002 Canadian Conference on Electrical and Computer Engineering Conference Proceedings (Cat No 02CH37373) CCECE-02, Winnipeg, MB, Canada, 12–15 May 2003; Volume 3, pp. 1583–1588.

22. Rhee, I.; Warrier, A.; Min, J.; Xu, L. DRAND: Distributed Randomized TDMA Scheduling for Wireless Ad Hoc Networks. *IEEE Trans. Mob. Comput.* **2009**, *8*, 1384–1396. [CrossRef]

23. Ramanathan, S. A unified framework and algorithm for (T/F/C)DMA channel assignment in wireless networks. In Proceedings of the Proceedings of INFOCOM '97, Kobe, Japan, 7–11 April 2002; Volume 2, pp. 900–907.

24. Wang, Y.; Henning, I. A Deterministic Distributed TDMA Scheduling Algorithm for Wireless Sensor Networks. In Proceedings of the 2007 International Conference on Wireless Communications, Networking and Mobile Computing, Shanghai, China, 21–25 September 2007; pp. 2759–2762.

25. Vergados, D.J.; Amelina, N.; Jiang, Y.; Kralevska, K.; Granichin, O. Toward Optimal Distributed Node Scheduling in a Multihop Wireless Network Through Local Voting. *IEEE Trans. Wirel. Commun.* **2018**, *17*, 400–414. [CrossRef]

26. Lutz, J.; Colbourn, C.J.; Syrotiuk, V.R. ATLAS: Adaptive Topology- and Load-Aware Scheduling. *IEEE Trans. Mob. Comput.* **2013**, *13*, 2255–2268. [CrossRef]

27. Vergados, D.J.; Sgora, A.; Vergados, D.D.; Vouyioukas, D.; Anagnostopoulos, I. Fair TDMA scheduling in wireless multihop networks. *Telecommun. Syst.* **2010**, *50*, 181–198. [CrossRef]

28. Li, Y.; Zhang, X.; Zeng, J.; Wan, Y.; Ma, F. A Distributed TDMA Scheduling Algorithm Based on Energy-Topology Factor in Internet of Things. *IEEE Access* **2017**, *5*, 10757–10768. [CrossRef]

29. Ambigavathi, M.; Sridharan, D.; M, A. Energy efficient and load balanced priority queue algorithm for Wireless Body Area Network. *Futur. Gener. Comput. Syst.* **2018**, *88*, 586–593. [CrossRef]

30. Karim, L.; Nasser, N.; Taleb, T.; Alqallaf, A. An efficient priority packet scheduling algorithm for Wireless Sensor Network. In Proceedings of the 2012 IEEE International Conference on Communications (ICC), Ottawa, ON, Canada, 10–15 June 2012; pp. 334–338.

31. Li, X.; Chen, N.; Zhu, C.; Pei, C. Improved Efficient Priority-and-Activity-Based QoS MAC Protocol. In Proceedings of the 2013 5th International Conference on Intelligent Networking and Collaborative Systems, Xi'an, China, 9–11 September 2013; pp. 315–318.

32. Slama, I.; Jouaber, B.; Zeghlache, D. A Free Collision and Distributed Slot Assignment Algorithm for Wireless Sensor Networks. In Proceedings of the IEEE GLOBECOM 2008 IEEE Global Telecommunications Conference, New Orleans, LO, USA, 30 November–4 December 2008; pp. 1–6.
33. Liu, Y.; Elhanany, I.; Qi, H. An energy-efficient QoS-aware media access control protocol for wireless sensor networks. In Proceedings of the IEEE International Conference on Mobile Adhoc and Sensor Systems Conference, Washington, DC, USA, 7 November 2005; pp. 189–191.
34. Yigitel, M.A.; Incel, O.D.; Ersoy, C. QoS-aware MAC protocols for wireless sensor networks: A survey. *Comput. Netw.* **2011**, *55*, 1982–2004. [CrossRef]
35. Github. Available online: https://github.com/djvergad/local_voting (accessed on 1 August 2017).

Article

EERS: Energy-Efficient Reference Node Selection Algorithm for Synchronization in Industrial Wireless Sensor Networks

Mahmoud Elsharief [1], Mohamed A. Abd El-Gawad [1], Haneul Ko [2] and Sangheon Pack [1,*]

1   School of Electrical Engineering, Korea University, Seoul 02841, Korea; mahmoud@korea.ac.kr (M.E.); mgawad@korea.ac.kr (M.A.A.E.-G.)
2   Department of Computer Convergence Software, Korea University, Sejong 30019, Korea; heko@korea.ac.kr
*   Correspondence: shpack@korea.ac.kr

Received: 30 June 2020; Accepted: 20 July 2020; Published: 23 July 2020

Abstract: Time synchronization is an essential issue in industrial wireless sensor networks (IWSNs). It assists perfect coordinated communications among the sensor nodes to preserve battery power. Generally, time synchronization in IWSNs has two major aspects of energy consumption and accuracy. In the literature, the energy consumption has not received much attention in contrast to the accuracy. In this paper, focusing on the energy consumption aspect, we introduce an energy-efficient reference node selection (EERS) algorithm for time synchronization in IWSNs. It selects and schedules a minimal sequence of connected reference nodes that are responsible for spreading timing messages. EERS achieves energy consumption synchronization by reducing the number of transmitted messages among the sensor nodes. To evaluate the performance of EERS, we conducted extensive experiments with Arduino Nano RF sensors and revealed that EERS achieves considerably fewer messages than previous techniques, robust time synchronization (R-Sync), fast scheduling and accurate drift compensation for time synchronization (FADS), and low power scheduling for time synchronization protocols (LPSS). In addition, simulation results for a large sensor network of 450 nodes demonstrate that EERS reduces the whole number of transmitted messages by 52%, 30%, and 13% compared to R-Sync, FADS, and LPSS, respectively.

Keywords: energy-efficiency; industrial wireless sensor networks; reference node selection; time synchronization

---

1. Introduction

Industrial wireless sensor networks (IWSNs) have been widely developed in the last decade. They play a significant role in many applications, e.g., industry, healthcare, agriculture, and smart metering systems. Currently, IWSNs perform as an essential foundation in the widespread industrial internet of things [1,2]. IWSNs typically consist of many sensor nodes that are spatially disseminated over a region of interest. Generally, sensor nodes are usually used to observe environmental conditions [3]. They are often battery-operated, and sometimes it is infeasible to recharge or replace batteries. Consequently, the lifetime of the battery becomes a crucial concern in the design of IWSNs [4,5]. To maximize the battery lifetime and conserve the battery power, a time-synchronous operation is regularly preferred. Most of the sensor nodes need time synchronization to coordinate wake up and sleep operations at an arranged time. Consequently, sensor nodes require trust and robust synchronization [5].

Time synchronization is considered as a critical part of the current industrial operation of sensor nodes. It offers a common reference time for whole sensor nodes in the network. Generally, sensor nodes are frequently prepared with hardware clock oscillator, which is relatively inexpensive, and inaccurate.

In this way, it is much challenging to introduce precise synchronization for such sensor nodes [6,7]. With such sensor nodes, it is required to periodically conduct an accurate synchronization process to well-preserved time synchronization of networks for a long time [8]. Lack of synchronization in any sensor node can result in an imprecise wake-up time of the sensor node, which leads to a severe failure of network connectivity [9]. All sensor nodes, therefore, require regularly exchange timing messages in order to keep the whole network synchronized. Exchanging timing messages minimizes the time offsets that produced by the clock drift of each node in the network [7,10]. Frequent synchronization, however, dramatically raises energy consumption as the transmission of messages typically corresponds to the most significant portion of energy consumption [10].

In the literature, several conventional [11–14] and advanced [15–17] protocols for the synchronization of sensor nodes have been developed. The main aim of these protocols is to provide synchronization to all sensor nodes in networks. Most of these protocols were intended to focus on the accuracy of synchronization as it is very important with less concern of communication overhead. Besides, these protocols, due to frequent collisions, suffer from message loss. Over the past few years, a few synchronization protocols [5,6,18,19] have aimed to address the energy consumption issue. In general, however, the consideration gotten for the aspect of energy consumption is modestly lower contrasted to that for the synchronization accuracy [15]. We can conclude the energy consumption is a big challenge in the time synchronization of IWSNs. This challenge motivates us to introduce an energy-efficient reference node selection (EERS) algorithm for synchronization. It can substantially reduce the energy consumption by decreasing the number of transmitted messages among the nodes in the network. EERS minimizes the number of connected reference nodes and satisfies a minimum number of hops. Besides, EERS proposes a new method that avoids collisions completely among nodes during the time synchronization process. The contributions of EERS can be summarized as: (1) EERS significantly reduces the communication overhead (number of transmitting messages) and accelerates the reference scheduling, which can reduce energy consumption; (2) EERS employs a reference node scheduling method to resolve the collision problem and in turn to reduce power consumption; and (3) we implemented EERS in a real wireless sensor network with Arduino Nano RF sensors and conducted extensive large-scale simulations.

The remainder of this paper is organized as follows. We summarize the related works in Section 2. In Section 3, we elaborate on the system model and the operation of the EERS algorithm. In Section 4, we present the experimental and simulation results. The concluding remarks are finally given in Section 5.

2. Related Work

Time synchronization for sensor nodes has been considerably analyzed for decades. Within the literature, there have been many methods for clock synchronization that are aimed toward minimizing energy consumption via decreasing the communication overhead and increasing the accuracy. In this section, we focus on those works which are concerned with reducing the communication overhead by decreasing the number of transmitted messages.

Noh et al. proposed the pairwise broadcast synchronization (PBS) [20]. It utilizes the pairwise operation introduced in [21]. PBS reduces the energy consumption (number of transmitted messages) by employing an overhearing technique in a single hop domain. The authors extended their work by introducing the Multi-hop PBS called groupwise pair selection algorithm [22]. It consists of a pair selection methods and hierarchy forming. GPA, however, needs an additional pairwise operation that reduces the energy efficiency compared with its single-hop counterpart [7]. Selecting pairs of nodes in a large network can be very costly in terms of both computation time and energy.

Spanning tree-based energy-efficient time synchronization (STETS) [23] has been proposed for the industrial Internet of things (IIoT). It employs the sender-to-receiver protocol (SRP) as well as receiver-to-receiver protocol (RRP). Especially in a large-scale and densely connected network, it effectively decreases energy consumption by reducing the number of transmitted messages. STETS, however, cannot synchronize all the nodes in some cases. The robust time synchronization (R-sync) [6] for IIoT has been introduced to resolve the drawbacks of STETS. Similar to STETS, R-Sync utilizes SRP and RRP. It focuses on identifying and pulling back the isolated nodes to the network that have lost their synchronization. Although R-sync requires relatively fewer numbers of messages compared to the timing-sync protocol for sensor networks (TPSN) [14] and STETS, it does not have any technique to avoid collisions.

The energy-efficient coefficient exchange synchronization protocol (CESP) [19] was introduced by Gong et al. to address the excessive power consumption of the reference broadcast synchronization (RBS) [13]. CESP employs the synchronization coefficient to decrease the number of transmitted messages compared with RBS [10]. CESP, however, consumes relatively high power and does not have any technique to avoid collisions. Yildirim et al. proposed an adaptive value tracking synchronization (AVTS) [24] to resolve the drawback of quick flooding. AVTS provides a scalable synchronization and reduces memory overhead compared to the flooding time synchronization protocol (FTSP) [12] and flooding with clock speed agreement (FCSA) [25]. AVTS, however, uses the scheme of flooding messages which results in the low energy efficiency of the network.

The density table-based synchronization (DTSync) protocol [5] has been introduced. It utilizes the concept reference scheduling technique. Iteratively, the reference scheduling mechanism selects an ordered set of common nodes. These nodes are responsible for disseminating the timing messages in the entire network. Compared to hierarchy reference broadcast synchronization (HRTS) [11], DTSync requires fewer messages, which in turn minimize the energy consumption. Elsharief et al. introduced the fast scheduling and accurate drift compensation for time synchronization (FADS) [18]. Similar to DTSync, FADS employs a reference scheduling technique that organizes the message transmission among the sensor nodes. It improves the performance of the reference scheduling process by reducing the time consumption. FADS, however, during the reference scheduling process, produces a relatively large number of message broadcasts which causes excessive energy consumption [10].

Recently, we proposed a low power scheduling for time synchronization protocols (LPSS) [10]. Compared to FADS, LPSS can significantly reduce the number of broadcasted messages and accelerates the reference scheduling process in a centralized manner, which in turn reduces the energy consumption. Besides, it provides a scheme to avoid collisions. LPSS, however, selects reference nodes randomly at each level which leads to the inefficient energy usage of the network.

3. EERS Algorithm

In this section, we first describe the system model for the EERS algorithm and detail the operation of the EERS algorithm.

3.1. System Model

In this paper, a wireless sensor network is designed as an undirect graph $G = (V, E)$ consisting of a set of sensor nodes, $V = \{1, 2, \ldots., \ldots, N\}$, in the network and a set of links, $E \subseteq V \times V$, indicating the connection among the sensor nodes. That is, $(i, j) \in E$ if node i and node j are located within the transmission range of each other [26]. In this paper, each node has a unique ID number and is supplied with a hardware clock oscillator. For simplicity, we assume that the network is stationary, and the transmission among nodes is reliable. Besides, it is assumed that all sensor nodes are identical (i.e., they have the same transmission range of R meter radius). Moreover, we assume that the sink node recognizes the position of every node in the network.

3.2. Operation of EERS

EERS is a greedy heuristic algorithm that attempts to minimize the communication overhead. It guarantees the coverage of all nodes and accelerates the reference node scheduling in a tree-based network topology. Additionally, to guarantee collision avoidance during the synchronization process, EERS assigns an exclusive scheduled time slot to each reference node.

The pseudo-code of EERS is presented in Algorithms 1 and 2. Algorithm 1 is used by the sink node only. It describes the process of selecting reference nodes, *RefNodes*, and their scheduled time slots, *SchSlots*. On the other hand, Algorithm 2 is used by other nodes in the network, which shows the procedure of receiving the *RefNodes* and *SchSlots*.

In the beginning, in Algorithm 1, to guarantee the minimum hops starting from sink node to farthest node in the network, sink node has perfect awareness of the network topology and utilizes the breadth-first search (BFS) [27] algorithm to determine the level of every node in the network (Algorithm 1, Line 5). As a result of BFS, we have a tree-based network consisting of L levels. Each level, k, has a set of nodes, N_k. The optimal aim is to find the minimum number of connected reference nodes. In fact, finding a minimum number of reference nodes is a non-deterministic polynomial-time (NP)-hard problem [4,26]. Specifically, its time complexity is $O(m \times m!) \leq O(m^m)$, where m is the number of nodes in each level. In this article, our target is finding a simple approximation algorithm to find the minimum set of reference nodes, $R_k \subseteq N_k$, that cover all nodes in the next level, $k + 1$. At the end of the initial step, the following steps are sequentially conducted.

Step 1: As the sink node, S, knows the position of each node, it determines the distance between each node in the level k and its neighbor nodes in the level $k + 1$ (Algorithm 1, Lines 7–9). The distance, $d_{i_k, \, j_{k+1}}$, between node $i_k \in N_k$, and node $j_{k+1} \in N_{k+1}$ is calculated as follows

$$d_{i_k, \, j_{k+1}} = \sqrt{(x_{i_k} - x_{j_{k+1}})^2 + (y_{i_k} - y_{j_{k+1}})^2} \tag{1}$$

where (x_{i_k}, y_{i_k}) and $(x_{j_{k+1}}, \, y_{j_{k+1}})$ are the position of the nodes i_k, and j_{k+1}, respectively. As a result of this process, EERS builds a set of the distances between N_k and N_{k+1}, $d_{N_{k+1}}^{N_k}$.

Algorithm 1: EERS pseudo-code for reference node selection (sink node only)

Initial :, *CoverNodesFlag* $\leftarrow$ *false*, *cnt* $\leftarrow$ 0, $k \leftarrow$ 0, *SeqN* $\leftarrow$ 0

1.	$s \leftarrow$ *sink node*
2.	*RefNodes*(1) $\leftarrow$ *s*
3.	*Schslot*(1) $\leftarrow$ 0
4.	*CoverNodesFlag*(*s*) $\leftarrow$ *true*
5.	Compute the hop distance from s and the level to every node (i.e., using BFS)
6.	**while** ($k \, ! = LevelMax - 1$) **do** // *LevelMax* is equal to the level of the farthest node from sink node
7.	**for** (*every edge* $(i, j) \in E$, $i \in N_k, j \in N_{k+1}$) **do**
8.	detrmine $d_{i,j}$
9.	**end for**
10.	**while** (*CoverNodesFlag*(N_{k+1}) $! = true$) **do**
11.	$R \leftarrow N_k\big(\max(d_{i,j})\big)$
12.	*RefNodes*(*cnt*) $\leftarrow$ *R*
13.	*SchSlots*(*cnt*) $\leftarrow$ *cnt*
14.	*CoverNodesFlag*(*m*) $\leftarrow$ *true* // ($m \subseteq N_{k+1}$) are the neighbor nodes of R
15.	*cnt* $\leftarrow$ *cnt* + 1
16.	**end while**
17.	$k \leftarrow k + 1$
18.	**end while**
19.	*SeqN* $\leftarrow$ *SeqN* + 1
20.	brodcast refernces scheduling message <*RefNodes, SchSlots, SeqN*>

Step 2: S selects the reference node, R using the equation (Algorithm 1, Line 11)

$$R = N_k\Big(Max\big(d_{N_{k+1}}^{N_k}\big)\Big) \tag{2}$$

where the function $Max(d_{N_{k+1}}^{N_k})$ chooses a node that can cover the farthest node in the $k + 1$ level.

Then, EERS adds R to *RefNodes*. Next, S assigns a scheduled time slot to R and adds it to *SchSlots* (Algorithm 1, Lines 12 and 13).

Step 3: To avoid processing a node in N_{k+1} more than once, we mark it as a covered node (Algorithm 1, Line 14). Then, we update the $d_{N_{k+1}}^{N_k}$ by removing the distance between it and other nodes in N_k.

Step 4: S increments the sequence number, $SeqN$ (Algorithm 1, Line 19). $SeqN$ is added to prevent sending the scheduling message many times. Next, S broadcasts the references scheduling message among all nodes in the networks (Algorithm 1, Line 20).

Steps 1–4 in Algorithm 1 are repeated until no more reference node can be chosen in the network. We can conclude the time complexity of EERS is $\sum_{k=0}^{LevelMax-1} O(N_k) \approx O(N)$ where *LevelMax* is the highest level in the network.

Upon receiving the scheduling message (Algorithm 2), each node only retracts if the received sequence number is larger than the stored one (Algorithm 2, Line 2). Next, the node checks if its ID has been recorded in *RefNodes*l then, it keeps the corresponding time slot, *mySchslot* (Algorithm 2, Lines 5–7). Next, every reference node initiates the waiting timer (WT). The value of WT can be computed as

---

Algorithm 2: EERS pseudo-code for reference scheduling selection (except sink node).

Initial : $cnt \leftarrow 0$, $SeqN \leftarrow 0$
1.   ■ **Upon receiving Scheduling Message**
2.   **if** $(Received\ SeqN > SeqN)$ **Then**
3. $SeqN \leftarrow Received\ SeqN$
4.   **for** $(cnt \leftarrow 0\ to\ length(RefNodes))$**do**
5.   **if** $(NodeID == RefNodes(cnt))$ **Then**
6. $mySchsSlot \leftarrow SchSlots(cnt);$
7.   **end if**
8.   **if** $(RefNodes(cnt) == SenderID)$ **Then**
9. $SchSlot_{SrcID} \leftarrow SchSlots(cnt)$
10.   **end if**
11.   **end for**
12. $WaitingTime \leftarrow (mySchSlot - SchSlot_{SrcID})$
13.   Setup waiting timer $WT(WaitingTime)$
14.   **end if**
15.   ■ **Upon WT expires**
16.   forward refernces scheduling message $<RefNodes, SchSlots, SeqN. >$

$$WTvalue = mySchslot - Schslot_{SrcID} \qquad (3)$$

where *WTvalue* is the time that each reference should wait before forwarding the scheduling message, $Schslot_{SrcID}$ is the sender's time slot, and *mySchslot* is the current reference time slot (Algorithm 2, Line 12). As soon as WT expires, the current reference forwards the scheduling message to its neighbor nodes (Algorithm 2, Lines 15 and 16).

Figure 1 shows an example of EERS. Node S applies Algorithm 1 to select reference nodes (see Figure 1a) and assigns an exclusive scheduled time slot to each reference (see Figure 1b). Upon receiving the reference scheduling message, each node applies Algorithm 2 to verify if it has been selected as a reference node. For example, in Figure 1, it is assumed that node F is selected as a reference node and its assigned scheduled time slot is 4. As soon as it receives a packet form node C, it determines its waiting time equals to 2 (=4 (i.e., F's *mySchSlot*) – 2 (i.e., $Schslot_{SrcID}$ of node C)). Then, F sets up its waiting timer and it forwards the received message to its neighbor nodes upon the WT is expired.

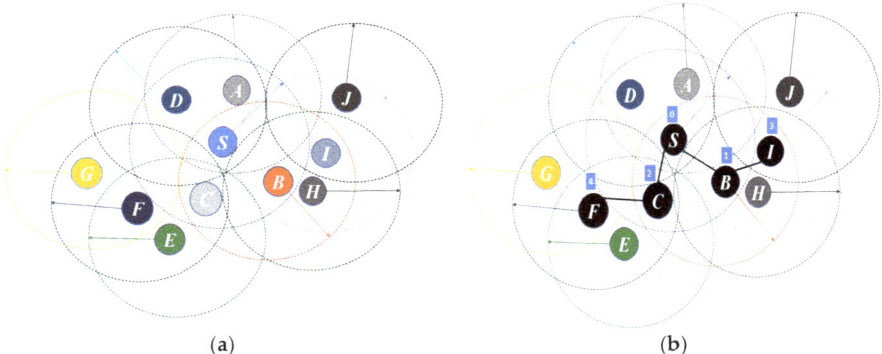

(a)   (b)

Figure 1. Wireless sensor network example. (**a**) A sensor network before applying EERS. (**b**) The EERS chooses the reference nodes (colored in black) and assigns schedule time slot to each reference.

4. Experimental and Simulation Results

In this section, we present experimental results achieved using a network of real hardware sensors. EERS was implemented in a C++ program and ported on an embedded processor in the hardware sensors. Next, we performed experiments with wireless networks composed of each hardware sensor running the C++ program. In addition, we compared our algorithm with the previous methods, FADS [18], LPSS [10], and R-Sync [6]. Note that the synchronization protocol in [5] was used for EERS to highlight the benefit of EERS.

To show the performance of EERS for a large-scale network, we also conducted a simulation study and compared the performance of EERS with FADS, LPSS, and R-Sync. In our experimental and simulation studies, the number of transmitted messages was quantitatively analyzed, and the time of the whole procedure was measured, as they have a significant impact on energy consumption. Generally, the wireless network is a broadcast domain. For each transmission, therefore, there are multiple receptions [28]. For example, assume that we have a uniformly distributed network with N nodes, and each node has B surrounding nodes. The total energy consumption of the transceiver, E_T, can be expressed by transmission energy $(E_{TX} = A.P_{TX}.T_{TX})$, and reception energy $(E_{RX} = A.B.P_{RX}.T_{RX})$ as defined by the following equation.

$$E_T = A(P_{TX}.T_{TX} + B.P_{RX}.T_{RX}) \tag{4}$$

where A is the number of transmitted messages, B is the number of received messages, P_{TX} is the transmission power, P_{RX} is the reception power, T_{TX} is the time duration of the transmitted packet, and T_{RX} is the time duration of the received packet. For the sake of simplicity, consider $P_{TX} \approx P_{TX} = P$, the packet length, L, and data rate, D, are fixed. Hence, $T_{TX} = T_{RX} = \frac{L}{D}$. We can, therefore, simplify Equation (4) as Equation (5).

$$E_T = A\frac{P.L(B+1)}{D} \tag{5}$$

From Equation (5), it can be noted that the energy consumption is directly proportional to the number of transmitted messages. Besides, reducing the number of messages decreases the time it takes for the transmitter node to transmit all its messages. This in turn allows the node to spend less time in idle mode while processing the transmission procedure, and to switch to sleep earlier. Since the idle mode also consumes energy [19], minimizing idle time can further save energy. Considering the transceiver of Arduino Nano RF [29], we conducted a wireless network of 25 nodes. Each node in the network can communicate with each other. Therefore, the total energy consumption per each transmission is equal to $\frac{49 \times 8 \times (33.9m + 24 \times 36.9m)}{250k} \approx 1.44\ mJ$.

4.1. Experiments Setting

In general, choosing a sensor module is dependent on the application's requirements (computing power, power consumption, memory, dedicated range, data rate, etc.) and the budget. Here, in this paper, our target is spreading the timing messages over a multi-hop network for smart metering applications. We chose an Arduino Nano RF [29] to implement the hardware for sensor nodes. The hardware integrates an ATmega328P CPU core [30] and an NRF24L01 Rf transceiver on-chip [31]. Arduino Nano RF is a cheap module with acceptable memory, data rate, and transmission range. However, EERS can be implemented on other types of sensors modules. Table 1 provides the specification of the sensor board. A photo of the sensor board is shown in Figure 2a. We adopted a wireless sensor network of 25 nodes, as displayed in Figure 2b. We configured a network of the four-way grid topology illustrated in Figure 3 to test EERS over a multi-hop network. For simplicity of the testing, all the nodes were configured with a preconfigured set of neighbor nodes and the distances among them. The solid lines in Figure 3 indicate such neighbor nodes assigned to every node. The sensor nodes can communicate with each other if they have a common solid line. With this experiment, the performance of EERS could be evaluated under realistic conditions of congestion, transmission contention, and various packet collisions. Even though we did the tests using a simple network of small size in the laboratory, these tests demonstrated that EERS can scale to networks of a large scale with longer wireless range and can provide equally great execution.

Table 1. Specification of sensor module.

Parameter	Value
Carrier frequency	2.4 GHz
Memory	2 KB RAM, 32 KB Flash
Bit rate	250 Kbps
Transmission power	33.9 mw
Reception power	36.9 mw
Modulation type	GFSK
TX power	0 dBm
Frame size	49 Bytes
Scheduled time slot	10 ms
Crystal oscillator frequency	16 MHz

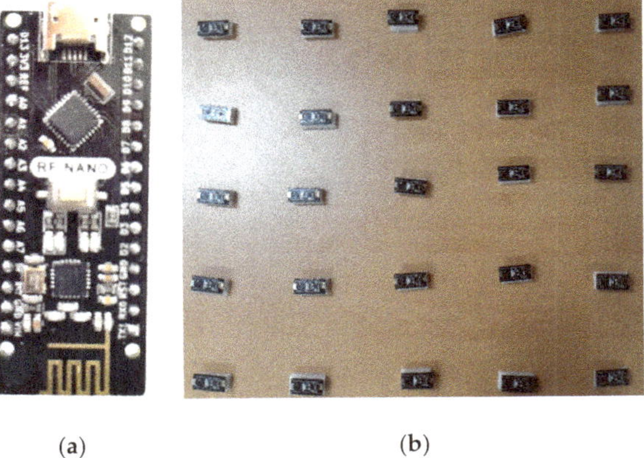

(a)   (b)

Figure 2. (a) Hardware platform (b) Set up of a network of 25 sensor nodes for the experiment.

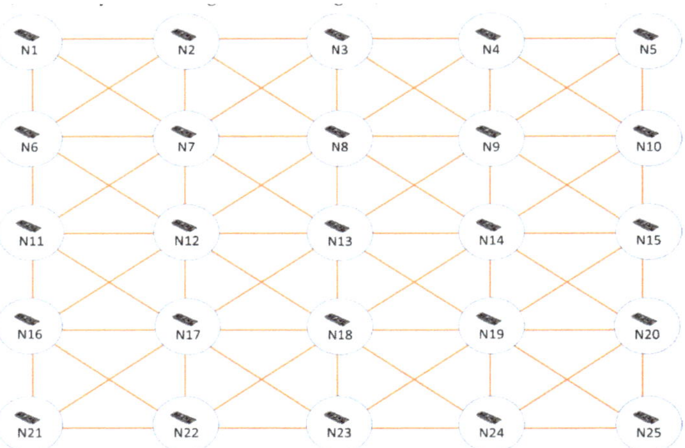

Figure 3. Sensor node in 4-way grid topology.

4.2. Experimental Results

This section describes the evaluation of the EERS, LPSS, FADS, and R-sync protocols. First, we evaluate the reference node scheduling process. R-sync is excluded from this evaluation as it does not have any reference scheduling mechanism. In the reference node scheduling procedure, the key performance indicator is the number of transmitted messages and the consumed time of the whole process. Table 2 shows that EERS has considerably fewer messages than LPSS and FADS for the scheduling process. For the network topology of Figure 3, EERS sent only 11 messages, while LPSS sent 17 messages (around 1.5 times more) and FADS exchanged 24 messages (around two times more). Next, for time consumed by the reference node scheduling procedure, the experimental results show EERS is faster than that of LPSS and FADS. Table 2 also compares the processing time measured for EERS, LPSS, and FADS protocols. EERS can reduce the whole time of the process to around 35% and 76% compared to LPSS and FADS, respectively.

Table 2. Comparison of EERS, LPSS, FADS and R-sync.

Protocol Operation	EERS	LPSS	FADS	R-Sync
Scheduling process (message)	11	17	25	-
Time of scheduling process (second)	1.1	1.7	46	-
Synchronization process (message)	27	45	27	51

Finally, we applied the synchronization protocol of [5] on top of the EERS. The experimental results show that EERS and FADS transmit fewer messages than other protocols for the synchronization process. Table 2 shows that EERS has considerably fewer messages than LPSS and R-Sync. EERS and FADS required around 27 messages, while LPSS exchanged 45 messages (around 1.66 times more) and R-Sync exchanged 51 messages (around 1.88 times more).

4.3. Large-Scale Simulation Results

Experimental results reveal that EERS improves energy consumption in a small test network. To further validate EERS, we developed a simulator that conducts the algorithms of EERS, LPSS, and FADS. Meanwhile, the simulation results of R-Sync were adopted from [6]. Simulations were developed in MATLAB using wireless networks of various sizes, and different transmission ranges. We presumed there are N sensor nodes which are randomly distributed in a square area of

1000 m × 1000 m. We also assumed that all nodes are identical and independent. Key simulation parameters are summarized in Table 3. In fact, we adopted most of the simulation parameters from [6]. We believe the simulation parameters is an appropriate choice to simulate realistic scenarios. By variation of the transmission range, we could simulate many types of sensor nodes which have different wireless ranges. For example, Arduino Nano RF has up to 100 m radius, which is located between 85 and 160 m (simulation range). On the other hand, changing the number of nodes in the network allowed us to demonstrate the behavior of our protocol in the sparse network (200 nodes) up to the dense network (450 nodes). We focused on the number of transmitted messages and the time of the process to show the performance of EERS. To reduce the impact of random errors, we ran 10,000 cycles for each experiment and obtained their average value. To illustrate the effect of the position of the sink node on the simulation results, two positions of sink node were tested: (1) sink node located at the center of the network; and (2) sink node located at the corner of the network. As expected, there is no big difference in the results due to the uniform distribution of the nodes. Therefore, to avoid redundancy, we only report the results in which the sink node was located at the center of the network.

Table 3. Simulation Parameters.

Parameter	Value
Network area	1000 m ×1000 m
Networks size (N)	200, 250, 300, 350, 400 and 450 nodes
Nodes distribution	uniformly distribution
Node wireless range radius (r)	85, 100, 115, 130, 145, and 160 m
Scheduling time width	10ms
Sink node position	Center (500 m, 500 m), Corner (0,0)

First, we evaluate the performance of all protocols (i.e., EERS, LPSS, FADS, and R-Sync) under transmission ranges from 85 to 160 m. Figure 4 shows the number of transmitted messages at the reference node scheduling process while Figure 5 shows the time of the whole process. For a network with the transmission range of 160 m, EERS, LPSS, and FADS generate 61, 71, and 380 messages, respectively. EERS outperforms LPSS and FADS by factors of 1.2 and 6.3, respectively. On the other hand, EER can decrease the time of the scheduling process to around 14% and 84% compared to LPSS and FADS, respectively, as shown in Figure 5. In the synchronization process. EERS has considerably fewer messages than LPSS, FADS, and R-Sync, as depicted in Figure 6. EERS outperforms LPSS, FADS, and R-Sync by factors of 1.17, 1.34, and 1.44, respectively. Next, we evaluate the performance of EERS, LPSS, FADS, and R-Sync over various network sizes from 200 to 450 nodes. Figure 7 shows the number of transmitted messages at the reference node scheduling process. For a large network with 450 nodes, EERS, LPSS, and FADS generate 157, 180, and 834 messages, respectively. EERS outperforms LPSS and FADS by factors of 1.15 and 5.3, respectively. On the other hand, EER can decrease the time of the scheduling process to around 13% and 81% compared to LPSS and FADS, respectively, as shown in Figure 8. In the synchronization process, for a large network with 450 nodes, EERS has considerably fewer messages than LPSS, FADS, and R-Sync, as depicted in Figure 9. EERS outperforms LPSS, FADS, and R-Sync by factors of 1.15, 1.4, and 2.1, respectively.

From the above results, it can be concluded that EERS consumes less energy than other techniques. At the same time, EERS does not lose the accuracy of the timing. Additionally, EERS is well-suited for those applications in which the position of each node is recognized, e.g., smart metering. Further, without losing its properties, EERS can handle high-density networks and therefore it can be considered as a highly scalable and adaptive protocol.

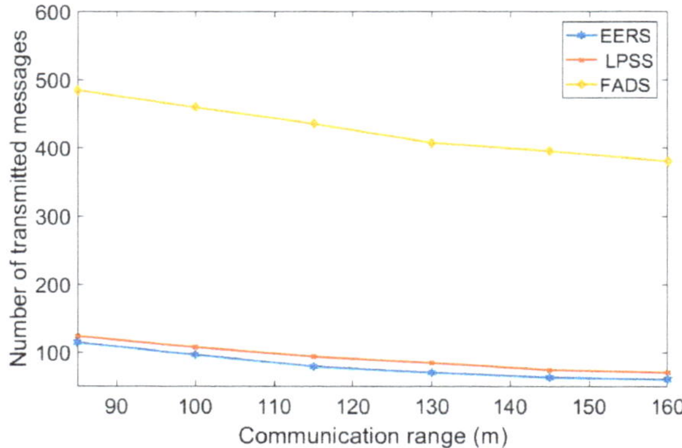

Figure 4. Required number of messages for reference node scheduling process (r = 85 m: 160 m, N = 240).

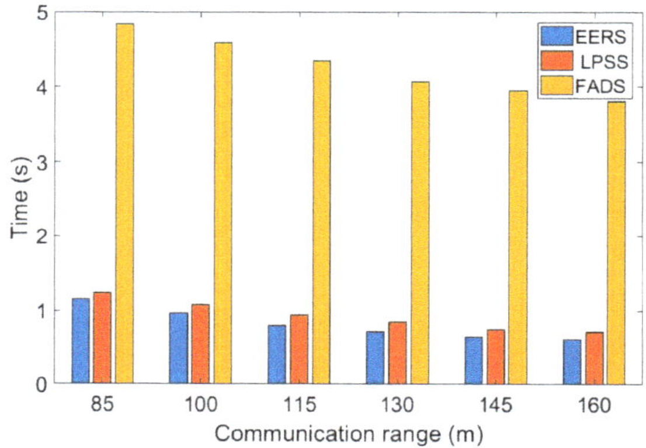

Figure 5. Required time for reference node scheduling process (r = 85 m: 160 m, N = 240).

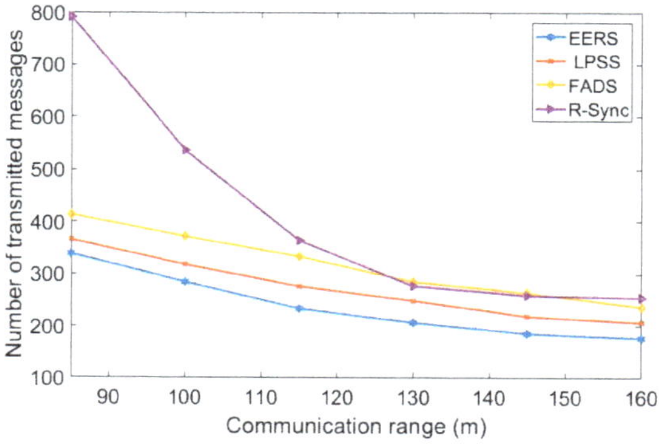

Figure 6. Required number of messages for synchronization process (r = 85 m: 160 m, N = 240).

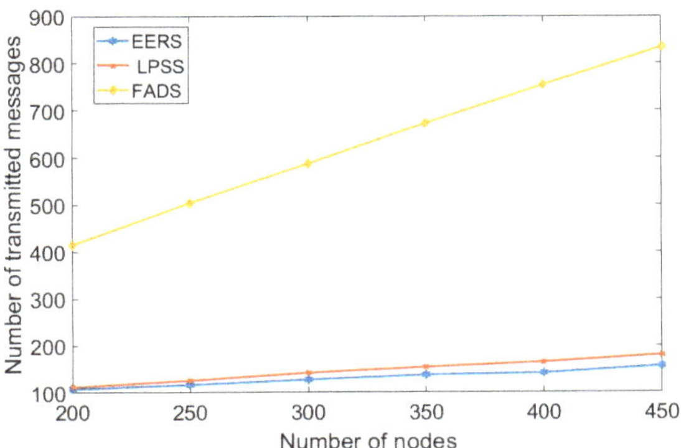

Figure 7. Required number of messages for reference node scheduling process (r = 85 m, N = 200: 450).

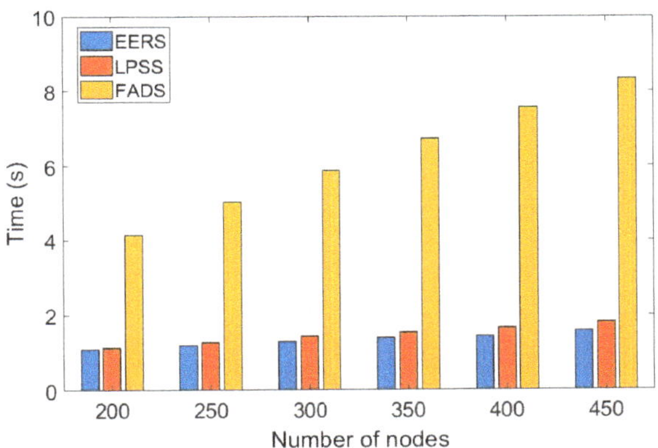

Figure 8. Required time for reference node scheduling process (r = 85 m, N = 200: 450).

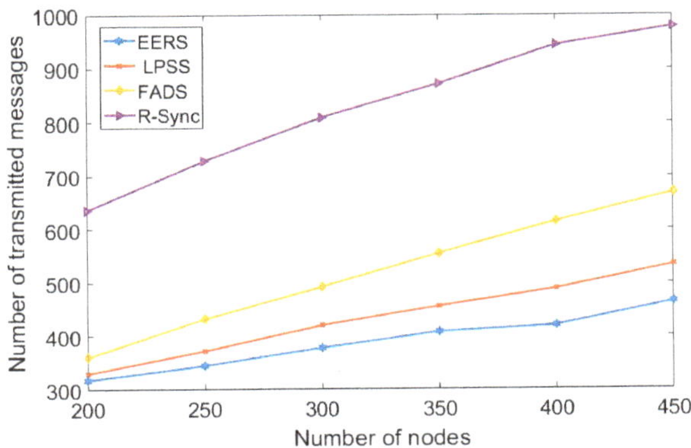

Figure 9. Required number of messages for synchronization process (r = 85 m, N = 200: 450).

5. Conclusions

In this paper, we consider the recent synchronization protocols for wireless sensor networks and introduce the energy-efficient reference node selection algorithm for synchronization in IWSNs (EERS).

EERS resolved major drawbacks of prior schemes (i.e., LPSS, FADS, and R-Sync) by introducing an efficient reference node scheduling process. EERS significantly reduced the number of messages in both scheduling and synchronization process, which leads to large savings in energy consumption. On the other hand, it accelerates the scheduling process. We conducted extensive simulations as well as the real measurement with hardware sensor nodes of the EERS algorithm. The experimental results show that EERS requires fewer messages than previous methods such as LPSS, FADS, and R-Sync.

As our future work, we are interested in combining deep learning techniques with EERS to find the optimal minimum set of reference nodes. The graph convolutional network will estimate the likelihood for each node in the graph to determine whether this node is part of the optimal solution [32]. Additionally, we plan to further improve EERS by allowing some of the reference nodes at the same level exchanging timing messages using the same scheduled time slot.

Author Contributions: M.E., M.A.A.E.-G. and S.P. created the idea of this research; M.E., H.K. and S.P. conceived the research methodology; M.E., M.A.A.E.-G. and H.K. performed the experiments and, simulation; and M.E., H.K. and S.P analyzed the data and wrote the article. All authors have read and agreed to the published version of the manuscript.

Funding: This research was supported by the National Research Foundation (NRF) of Korea Grant funded by the Korean Government (MSIP) (Nos. 2020R1A2C3006786 and 2019R1C1C1004352).

Acknowledgments: We gratefully appreciate the anonymous reviewers' variable reviews and comments.

Conflicts of Interest: The authors declare no conflict of interest.

References

1. Park, P.; Ergen, S.C.; Fischione, C.; Lu, C.; Johansson, K.H. Wireless Network Design for Control Systems: A Survey. *IEEE Commun. Surv. Tutor.* **2018**, *20*, 978–1013. [CrossRef]
2. Li, X.; Li, D.; Wan, J.; Vasilakos, A.V.; Lai, C.F.; Wang, S. A Review of Industrial Wireless Networks in the Context of Industry 4.0. *Wirel. Netw.* **2017**, *23*, 23–41. [CrossRef]
3. Gungor, V.C.; Hancke, G.P. Industrial Wireless Sensor Networks: Challenges, Design Principles, and Technical Approaches. *IEEE Trans. Ind. Electron.* **2009**, *56*, 4258–4265. [CrossRef]
4. Lisiecki, D.; Zhang, P.; Theel, O. CONE: A Connected Dominating Set-Based Flooding Protocol for Wireless Sensor Networks. *Sensors* **2019**, *19*, 2378. [CrossRef] [PubMed]
5. Elsharief, M.; Abd El-Gawad, M.A.; Kim, H. Density Table-Based Synchronization for Multi-Hop Wireless Sensor Networks. *IEEE Access* **2018**, *6*, 1940–1953. [CrossRef]
6. Qiu, T.; Zhang, Y.; Qiao, D.; Zhang, X.; Wymore, M.L.; Sangaiah, A.K. A Robust Time Synchronization Scheme for Industrial Internet of Things. *IEEE Trans. Ind. Inform.* **2018**, *14*, 3570–3580. [CrossRef]
7. Sarvghadi, M.A.; Wan, T.C. Message Passing Based Time Synchronization in Wireless Sensor Networks: A Survey. *Int. J. Distrib. Sens. Netw.* **2016**, *5*, 1–21. [CrossRef]
8. Yıldırım, K.S. Gradient Descent Algorithm Inspired Adaptive Time Synchronization in Wireless Sensor Networks. *IEEE Sens. J.* **2016**, *16*, 5463–5470. [CrossRef]
9. Silva, F.A. Industrial Wireless Sensor Networks: Applications, Protocols, and Standards [Book News]. *IEEE Ind. Electron. Mag.* **2014**, *8*, 67–68. [CrossRef]
10. Elsharief, M.; Abd El-Gawad, M.A.; Kim, H. Low-Power Scheduling for Time Synchronization Protocols in A Wireless Sensor Networks. *IEEE Sens. Lett.* **2019**, *3*, 1–4. [CrossRef]
11. Dai, H.; Han, R. TSync: A Lightweight Bidirectional Time Synchronization Service for Wireless Sensor Networks. *ACM SIGMOBILE Mob. Comput. Commun. Rev.* **2004**, *8*, 125–139. [CrossRef]
12. Maróti, M.; Kusy, B.; Simon, G.; Lédeczi, Á. The Flooding Time Synchronization Protocol. In Proceedings of the 2nd International Conference on Embedded Networked Sensor Systems, SenSys '04, Baltimore, MD, USA, 3–5 November 2004; pp. 39–49.
13. Elson, J.; Girod, L.; Estrin, D. Fine-Grained Network Time Synchronization Using Reference Broadcasts. *ACM SIGOPS Oper. Syst. Rev.* **2003**, *36*, 147–163. [CrossRef]
14. Ganeriwal, S.; Kumar, R.; Srivastava, M.B. Timing-Sync Protocol for Sensor Networks. In *Proceedings of the 1st International Conference on Embedded Networked Sensor Systems*; Association for Computing Machinery: New York, NY, USA, 2003.

15. Huan, X.; Kim, K.S.; Lee, S.; Lim, E.G.; Marshall, A. A Beaconless Asymmetric Energy-Efficient Time Synchronization Scheme for Resource-Constrained Multi-Hop Wireless Sensor Networks. *IEEE Trans. Commun.* **2020**, *68*, 1716–1730. [CrossRef]
16. Shi, F.; Tuo, X.; Yang, S.X.; Lu, J.; Li, H. Rapid-Flooding Time Synchronization for Large-Scale Wireless Sensor Networks. *IEEE Trans. Ind. Inform.* **2020**, *16*, 1581–1590. [CrossRef]
17. Elsts, A.; Fafoutis, X.; Duquennoy, S.; Oikonomou, G.; Piechocki, R.; Craddock, I. Temperature-Resilient Time Synchronization for the Internet of Things. *IEEE Trans. Ind. Inform.* **2018**, *14*, 2241–2250. [CrossRef]
18. Elsharief, M.; Abd El-Gawad, M.A.; Kim, H. FADS: Fast Scheduling and Accurate Drift Compensation for Time Synchronization of Wireless Sensor Networks. *IEEE Access* **2018**, *6*, 65507–65520. [CrossRef]
19. Gong, F.; Sichitiu, M.L. CESP: A Low-Power High-Accuracy Time Synchronization Protocol. *IEEE Trans. Veh. Technol.* **2016**, *65*, 2387–2396. [CrossRef]
20. Noh, K.L.; Serpedin, E.; Qaraqe, K. A New Approach for Time Synchronization in Wireless Sensor Networks: Pairwise Broadcast Synchronization. *IEEE Trans. Wirel. Commun.* **2008**, *7*, 3318–3322.
21. Noh, K.L.; Chaudhari, Q.M.; Serpedin, E.; Suter, B.W. Novel Clock Phase Offset and Skew Estimation Using Two-Way Timing Message Exchanges for Wireless Sensor Networks. *IEEE Trans. Commun.* **2007**, *55*, 766–777. [CrossRef]
22. Noh, K.L.; Wu, Y.C.; Qaraqe, K.; Suter, B.W. Extension of Pairwise Broadcast Clock Synchronization for Multicluster Sensor Networks. *EURASIP J. Adv. Signal. Process.* **2007**, *2008*, 286168. [CrossRef]
23. Qiu, T.; Chi, L.; Guo, W.; Zhang, Y. STETS: A Novel Energy-Efficient Time Synchronization Scheme Based on Embedded Networking Devices. *Microprocess. Microsyst.* **2015**, *39*, 1285–1295. [CrossRef]
24. Yıldırım, K.S.; Gürcan, Ö. Efficient Time Synchronization in a Wireless Sensor Network by Adaptive Value Tracking. *IEEE Trans. Wirel. Commun.* **2014**, *13*, 3650–3664. [CrossRef]
25. Yildirim, K.S.; Kantarci, A. Time Synchronization Based on Slow-Flooding in Wireless Sensor Networks. *IEEE Trans. Parallel Distrib. Syst.* **2014**, *25*, 244–253. [CrossRef]
26. Kim, D.; Wu, Y.; Li, Y.; Zou, F.; Du, D.Z. Constructing Minimum Connected Dominating Sets with Bounded Diameters in Wireless Networks. *IEEE Trans. Parallel Distrib. Syst.* **2009**, *20*, 147–157.
27. Cormen, T.H.; Leiserson, C.E.; Rivest, R.L.; Stein, C. *Introduction to Algorithms*; MIT Press: Cambridge, MA, USA, 2009.
28. IEEE Standard for Low-Rate Wireless Networks. IEEE Std 802.15.4-2015 (Revision of IEEE Std 802.15.4-2011). Available online: https://ieeexplore.ieee.org/document/7460875 (accessed on 11 April 2020).
29. Arduino Nano User Manual. Available online: https://www.arduino.cc/en/uploads/Main/ArduinoNanoManual23.pdf (accessed on 1 June 2020).
30. ATmega328P [DATASHEET]. Available online: http://ww1.microchip.com/downloads/en/DeviceDoc/Atmel-7810-Automotive-Microcontrollers-ATmega328P_Datasheet.pdf (accessed on 1 June 2020).
31. nRF24L01+ Preliminary Product Specification. Available online: https://www.sparkfun.com/datasheets/Components/SMD/nRF24L01Pluss_Preliminary_Product_Specification_v1_0.pdf (accessed on 1 June 2020).
32. Li, Z.; Chen, Q.; Koltun, V. Combinatorial Optimization with Graph Convolutional Networks and Guided Tree Search. *arXiv* **2018**, arXiv:1810.10659.

Article

Dynamic Reconfiguration of Cluster-Tree Wireless Sensor Networks to Handle Communication Overloads in Disaster-Related Situations

Miguel Lino [1,2], Erico Leão [1], André Soares [1], Carlos Montez [2], Francisco Vasques [3,*] and Ricardo Moraes [4]

[1] Department of Computing, Federal University of Piauí, Teresina 64049-550, Brazil; miguel.neto@posgrad.ufsc.br (M.L.); ericoleao@ufpi.edu.br (E.L.); andre.soares@ufpi.edu.br (A.S.)

[2] Department of Automation and Systems, Federal University of Santa Catarina, Florianópolis 88040-900, Brazil; carlos.montez@ufsc.br

[3] INEGI, Faculty of Engineering, University of Porto, 4200-465 Porto, Portugal;

[4] Department of Computing, Federal University of Santa Catarina, Araranguá 88905-120, Brazil; ricardo.moraes@ufsc.br

* Correspondence: vasques@fe.up.pt

Received: 24 July 2020; Accepted: 18 August 2020; Published: 20 August 2020

Abstract: The development of flexible and efficient communication mechanisms is of paramount importance within the context of the Internet of Things (IoT) paradigm. IoT has been used for industrial, commercial, and residential applications, and the IEEE 802.15.4/ZigBee standard is one of the most suitable protocols for this purpose. This protocol is now frequently used to implement large-scale Wireless Sensor Networks (WSNs). In industrial settings, it is becoming increasingly common to deploy cluster-tree WSNs, a complex IEEE 802.15.4/ZigBee-based peer-to-peer network topology, to monitor and control critical processes such as those related to oil or gas, mining, or certain specific chemicals. The remote monitoring of critical events for hazards or disaster detection in large areas is a challenging issue, since the occurrence of events in the monitored environment may severely stress the regular operation of the network. This paper proposes the *Dynamic REconfiguration mechanism of cluster-Tree WSNs* (DyRET), which is able to dynamically reconfigure large-scale IEEE 802.15.4 cluster-tree WSNs, and to assign communication resources to the overloaded branches of the tree based on the accumulated network load generated by each of the sensor nodes. A complete simulation assessment demonstrates the proposed mechanism's efficiency, and the results show that it can guarantee the required quality of service level for the dynamic reconfiguration of cluster-tree networks.

Keywords: disaster; hazards; cluster-tree; remote sensing; industrial wireless sensor network

---

1. Introduction

Industrial plants are often constructed on large industrial sites, and involve multiple mechanical or chemical processes that are sometimes deployed in risk-prone outdoor areas. The risks posed by natural hazards can be extensive, and this implies a need for uninterrupted monitoring of environmental variables and specific dangerous events that may occur.

Real-time data collection and remote monitoring of events over large areas is a challenging issue, and this is conventionally aided by satellite imaging applications that can facilitate the development of disaster detection applications, such as landslide hazard monitoring and fire or forest post-fire detection [1]. The recent development of numerous forms of sensors and the recent advances in wireless communication and micro-nano electronic devices have leveraged the use of WSNs for these types of monitoring applications. A WSN can offer several advantages, such as in situ sensing closer to the monitored data, online detection of events, and faster deployment of the monitoring infrastructure [2,3].

However, to ensure the success of this type of monitoring, several technical challenges need to be overcome. Large-scale monitoring applications generally require complex network topologies to achieve adequate spatial coverage at the same time as communication with low packet losses and low delays. WSN nodes impose an additional constraint in terms of an energy-saving mode of operation. Due to the large scale of the areas monitored in this way, and the possible existence of obstacles both in indoor and outdoor environments, the development of adequate communication mechanisms is a major focus of research in relation to this type of problem.

In the literature, several communication protocols and technologies have characteristics that make them candidates for large-scale monitoring applications, such as Lora [4], Sigfox [5], IEEE 802.15.4 [6], and ZigBee [7]. The first two of these are Low-Power Wide-Area Networks (LPWANs), which are suitable for long-range communication with meagre bit rates. In turn, the IEEE 802.15.4/ZigBee set of standards is a Low-Rate Wireless Personal Area Network (LR-WPAN), which has become the de facto communication method for WSNs.

As IEEE 802.15.4 radios are not intended for communication over long ranges, the use of adequate peer-to-peer communication mechanisms is required in order to allow for coverage of large areas. The IEEE 802.15.4 and ZigBee protocols support a hierarchical peer-to-peer topology called a cluster-tree [8], where each cluster consists of a group of sensor nodes coordinated by a particular node called the Cluster-Head (CH). In a conventional periodic monitoring operation, sensor nodes monitor the environment and send the acquired data to their CHs, which gather all data from within the cluster and send them towards a Base Station (BS). The main CH of the entire network assumes the role of the BS—a sink node that collects and processes packets sent by all sensor nodes. This type of communication, from all nodes to a central node, is called convergecast communication.

The adequate configuration of both beacon scheduling and other network parameters, such as the buffer sizes, Superframe Duration (SD), and Beacon Interval (BI), is a critical issue. Underprovision of network resources can cause packet losses, while overprovisioning, i.e., the presence of slack in the schedule and buffers, tends to unnecessarily increase end-to-end communication delays. Among the network parameters that need to be considered in the beacon scheduling computation are the periodicity of data acquisition at the sensor nodes and the number of levels at each branch in the cluster-tree (e.g. the number of parent and child clusters of each CH). The resources are then statically allocated to the CHs by assuming the maximum values for each packet flow in each CH. However, the network behaviour may dynamically change over time, and this introduces several challenges that are not often addressed in existing proposals.

Disaster monitoring applications are inherently event-triggered; that is, the detection of measured values above a certain threshold can lead to the modification of the operational mode of the network in some regions of the network. For example, in an Industrial WSN (IWSN) fire risk detection application, the detection of high values for temperature in conjunction with low humidity can trigger an increase of the monitoring periodicity within the nodes located in that critical region. This modification will mean that the entire tree branch will need to be reconfigured to prioritise these particular packet flows; otherwise, data conveying critical information will suffer longer delays and/or will be discarded throughout the network.

This paper aims to demonstrate that a dynamic reconfiguration of the network must be performed in such cases since a static configuration implies the reservation of network resources for all CHs based on worst-case assessments. That is, maximum periodicities are assumed for all sensor nodes and the maximum number of packets is assumed to traverse each cluster. As a consequence, beacon scheduling may become unfeasible or, at least, the network may be overused, which will have severe consequences in terms of energy consumption. The reasoning behind this work is that there is a need to dynamically reschedule the network whenever there is a need to implement a change in the mode that the network works. The proposed DyRET communication mechanism addresses this requirement, and enables dynamic mode changes in cluster-tree networks by reallocating CH communication resources according to the needs of the supported applications. The use of DyRET allows for an initial configuration of the network based on the nominal load imposed by regular monitoring activities, and the reallocation of network resources on demand whenever necessary. For example, whenever a critical event occurs in the network, such as monitored data indicating the detection of a possible disaster situation, special attention needs to be paid to this region of the network, requiring its sensor nodes to increase their duty cycles. A reconfiguration of the operating parameters is required in order to guarantee that this critical event will not congest a whole branch of the network.

1.1. Objective and Contributions of This Paper

IWSNs must be able to deal with typical impairments in communication related to signal interference and the requirements for long lifetimes and reliable network operation [9]. These types of requirements are usually important when the monitored area is large. Although a cluster-tree is generally a suitable topology for WSNs when dealing with the monitoring of large areas, several technical issues must be carefully handled, such as setting up the scheduling of active cluster periods [10], efficient allocation of resources according to performance limitations [11], prioritising different types of data traffic, and dynamic reconfiguration of the overall network. The DyRET mechanism specifically addresses this last issue. The main contributions of this paper can be summarised as follows:

1. The proposal of a dynamic communication mechanism that is able to reconfigure large-scale cluster-tree IWSNs, triggered by the detection of critical events in the monitoring environment. This mechanism allows the PAN Coordinator to reconfigure the main parameters related to communication structures, such as BI and SD, with the aim of avoiding overload of the CH buffers and network congestion. It implements a mode change scheduling scheme for cluster-tree networks that is able to control and prioritize the traffic from specific message flows.
2. A simulation assessment of the proposed communication mechanism that considers different reconfiguration scenarios and network metrics.

1.2. Outline of This Work

The remainder of this paper is organised as follows. In Section 2 some background issues about IEEE 802.15.4/ZigBee and cluster-tree features are discussed. Related work is summarised in Section 3. Section 4 presents the problem statement of this proposal. Section 5 introduces the DyRET, a mechanism to dynamically reconfigure cluster-tree networks according to the occurrence of critical events in specific areas of the network. Section 6 presents the simulation assessment of the proposed reconfiguration mechanism and discusses the results. Finally, some conclusions and further considerations are presented in Section 7.

2. Ieee 802.15.4 and Zigbee

The industry digitalisation gave rise to the smart industry concept, also known as Industry 4.0. One of the factors that drove this digitalisation is the consolidation of technologies related to the IoT and the Industrial IoT (IIoT) paradigms [12,13], where wireless technology plays a fundamental role, providing appropriate support for the applications, offering advantages over wired technology in terms of flexibility, fast deployment, scalability, distributed processing capacity, and high mobility.

Within this context, the IEEE 802.15.4 and ZigBee set of standards is pointed out as the most widely used protocol stack for implementing WSNs. While the IEEE 802.15.4 presents the PHYsical layer (PHY) and Medium Access Control (MAC) sublayer specifications for LR-WPAN applications, ZigBee specifies the upper layers (Networks, Application and Security).

Basically, IEEE 802.15.4 standard defines two types of nodes: *Full Function Devices* (FFD) and *Reduced Function Devices* (RFD). FFDs can perform complex tasks, such as: routing, coordinating neighbour nodes, aggregation, fusion or filtering data, and physical sensing. RFDs are responsible only for sensing and transmitting physical data.

Depending on the type of application, IEEE 802.15.4/ZigBee standards support two basic types of network topologies: star and peer-to-peer. Unlike star WSNs, in which all sensor nodes are directly connected to the coordinator node (centralised communication paradigm), peer-to-peer networks can implement more complex topological formations, such as grid, mesh, and cluster-tree networks.

Cluster-tree is a special peer-to-peer network topology and is pointed out as one of the most suitable topologies to deploy large-scale WSNs [8]. In this topology, sensor nodes are grouped into neighbouring clusters, which are coordinated by CHs, as illustrated in Figure 1a. CHs are responsible for creating their own clusters and for synchronising the communication with their child nodes.

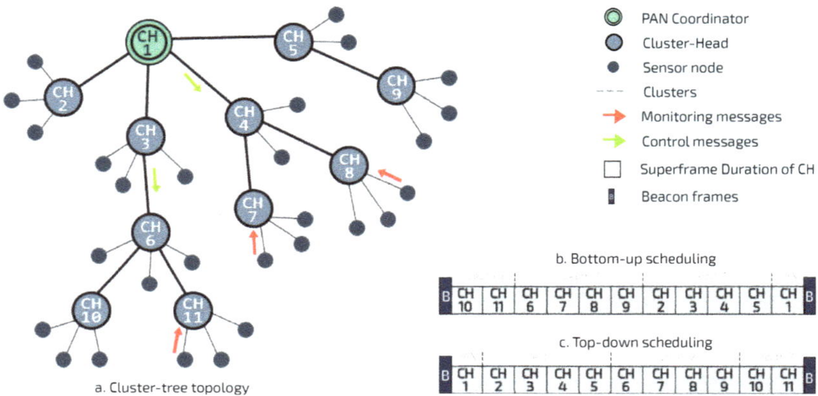

Figure 1. The cluster-tree WSN characteristics and the types of scheduling of its different traffic.

CH nodes are interconnected by parent-child relationships, forming a hierarchical structure that allows greater scalability than star networks. In this way, the cluster-tree routing is deterministic, following the tree levels (depths). In cluster-tree networks, the BS is often the coordinator of the Personal Area Network (PAN), i.e., the first and main CH of the network is the root node. This node is responsible for network management. Each CH synchronises its communication period with that of the PAN coordinator via beacon frame exchanges; and the PAN coordinator is responsible for organising the scale of beacon sending for the whole network.

The cluster-tree network operates in beacon-enabled mode, where beacon frames are used to synchronise the sensor nodes and they define a communication structure called superframe, illustrated in Figure 2. Superframes are delimited by beacon frames, that are periodically transmitted by all CHs (included PAN coordinator).

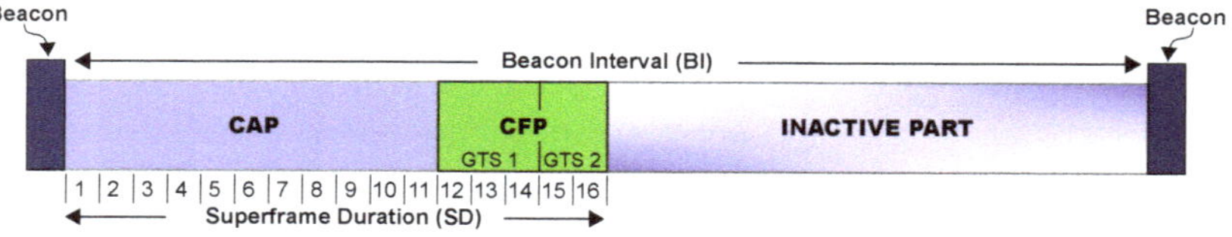

Figure 2. The superframe structure.

Basically, the superframe is defined by two parameters: *macBeaconOrder* (BO) and *macSuperframeOrder* (SO). These parameters define the Beacon Interval (BI) and the Superframe

Duration (SD), respectively. BI corresponds to the interval at which a cluster-head must periodically transmit its beacon frames. In turn, SD defines the period of communication of the clusters. The BI and SD are defined as follows:

$$BI = aBaseSuperframeDuration \times 2^{BO}, \qquad 0 \leq BO \leq 14$$
$$SD = aBaseSuperframeDuration \times 2^{SO}, \qquad 0 \leq SO \leq BO \leq 14 \tag{1}$$

where BO = 15 indicates that the network is operating in the non-beacon enabled mode. The *aBaseSuperframeDuration* corresponds to the minimum duration of a superframe when SO = 0 (by default, this parameter is equal to 960 symbols, corresponding to a duration of 15.36 ms, considering a bit rate of 250 kbps, frequency band of 2.4 GHz, and one symbol as 4 bits).

The beacon interval has an active part and, optionally, an inactive part. Thus, when BO is larger than SO, it means that exists an inactive part and sensor nodes can enter power save mode. When SO is equal to BO, there is no inactive part, i.e., the devices do not have additional time to save energy.

In the active part, the superframe starts immediately after the beacon frame, defining the period within which the nodes, both coordinators and sensors, can exchange messages. The active part is subdivided in two periods: Contention Access Period (CAP) and Contention Free Period (CFP). During CAP, sensor nodes compete to access the wireless channel using the *Carrier Sense Multiple Access-Collision Avoidance* (CSMA-CA) algorithm, as a form of collision avoidance. CFP is optional, and if requested, allows the CH to reserve Guaranteed Time Slots (GTS) so that a specific associated node has dedicated channel access and transmit contention-free messages.

From the point of view of communication mode, after the network formation, the data packets can be traveling *upstream* and *downstream*. Upstream traffic is the typical monitoring traffic, consisting of messages generated by sensor nodes that are forwarded to ascendant CHs until the PAN Coordinator. Reversely, downstream traffic corresponds to the traffic generated by the PAN Coordinator and forwarded to the descendent nodes.

There is no need for a clock synchronisation protocol to synchronise the sending of periodic beacons by neighboring CHs since the IEEE 802.15.4 MAC sublayer is responsible for this task. However, in order to avoid intercluster interferences and collisions of beacons and data frames, the active period of clusters must be organised. This is possible by applying beacon scheduling techniques, which correspond to the ordering of the transmission time to CHs' beacon frames. Basically, there are two types of beacon scheduling [10]: bottom-up and top-down, which respectively prioritise upstream and downstream traffic.

As outlined in Figure 1b, by using bottom-up scheduling, superframes are ordered following a bottom-up direction, where deepest clusters are firstly scheduled, depth-by-depth, until reaching the PAN coordinator. On the other hand, by using a top-down scheduling approach (Figure 1c), clusters are ordered from the PAN Coordinator, depth-by-depth, until reaching the deepest clusters.

3. Related Works

This section summarises the most relevant research works, addressing different issues: cluster scheduling [14,15], configuration of communication structures [10,11,16–19], data-load-based congestion control [20–27], and environmental monitoring network solutions for event-driven applications [28–30].

Regarding the beacon scheduling approaches, Koubaa et al. [14] summarise the problem of overlapping sensor nodes and highlight the risk of improperly configuring communication structures. The authors present different approaches to address direct and indirect collision issues. Firstly, the coordinator nodes transmit the beacon frames of all CHs early and, the other approach adjusts SDs of the same duration for simultaneous transmission.

In [15], a semi-dynamic scheduling scheme that allows non-coordinating nodes to act as CHs is proposed. These nodes can send data to the PAN Coordinator without waiting for the next actuation period. This is preceded by an algorithm that statically defines the beacon time and time slots for CH nodes and dynamically defines these features for all the sensor nodes. In addition, the time slot is assigned to the sensor node based on standard traffic and the availability verified by its CH according to node ID. These techniques are statically performed.

Regarding to configuration approaches of communication structures, Severino et al. [11] propose a cluster-tree designed to dynamically reorder CHs and reallocate their bandwidth. The reordering scheme (*Dynamic Cluster scheduling Reordering*—DCR) comprises an algorithm that performs scheduling based on the priority, number of cycles, neighbour set and depth of CHs. In turn, the allocation scheme (*Dynamic Bandwidth Re-allocation*—DBR) increases the bandwidth of CHs, whereas it reduces the bandwidth of others. However, this approach does not consider the load imposed by sensor nodes.

Kim and Kim [16] propose an energy-efficient reconfiguration algorithm that periodically selects CHs according to the shorter distance routes and lower energy cost whenever a threshold is reached. In contrast, the work presented in [17] builds a non-threshold cluster-head rotation scheme considering different energy resources (aggregation, transmission, residual and regular operations energy). As with [31], it also considers the depth and the load processed by each node. The mechanism proposed by Choudhury et al. [17] was compared to methods [18] based on the LEACH protocol [19], resulting in some gain in battery consumption and number of clusters, but it does not deal with the network congestion or random network formation issues.

To improve *Time Division Cluster Scheduling* (TDCS) algorithm [21], which deals with different directions of data flows, Ahmad and Hanzálek [22] propose a new heuristic method. In [21], it is proposed a method where messages between clusters are sent every period, considering a collision domain and based on the Integer Linear Programming theory for instances of small size (less than one hundred nodes). flows, Ahmad and Hanzálek [22] also propose the TDCS-PCC (Period Crossing Constraint), which deals with multiple collision domains, allowing messages in different streams to flow through better-defined paths based on graph heuristics, tree depths, and consecutive cluster paths. Despite these techniques to contribute to network life, event-driven large-scale applications are not addressed.

In recent years, a large-scale IWSN grouped into clusters for monitoring areas of toxic gas leakage was proposed by by Mukherjee et al. [28]. The main idea is to extend the lifetime of the network by activating the smallest number of nodes. In this approach, both the initial network formation and the selection of which nodes are activated is carried out using the *Connected K-Neighbourhood* (CKN) algorithm. The event is not considered, but the status of the zone is notified. In [29], a clustering and routing method for monitoring IWSN in fire-focused environments is presented. A hybrid CH selection scheme is implemented to benefit network energy efficiency. Then, the routing phase is adaptively configured as critical events are detected in the clusters. Events are reported using flags, but the data frame format is also changed.

Following the idea of event notification using data frames, the *Priority-based Congestion Control Protocol* (PCCP), proposed by [30], aims to prioritise upstream traffic flows according to three components: (1) *Intelligent Congestion Detection* (ICD); (2) *Implicit Congestion Notification* (ICN); and (3) *Priority-based Rate Adjustment* (PRA) hop-by-hop, in order to obtain weighted transfer rates among sensor nodes. While the ICD technique infers the existence of congestion by counting the number of packets sent locally, the ICN component is an efficient way of transporting congestion information piggybacking it in the header of data packets. Besides, the PRA method intends to allocate bandwidth based on the sensor nodes' priority, despite not defining which policy is used to assign the priorities.

The *Fairness Aware Congestion Control* (FACC) [20] implements a model of fairness bandwidth allocation in WSNs, by using a mechanism that divides the network in two categories: the aggregation nodes located near the *sink* node and the local acting nodes near the sensor nodes. FACC acts locally by regulating the rate of sensor nodes close to the coordinator node (*origin*) and acts globally by triggering reconfiguration messages from nodes near the *sink* node. When a packet is lost, the nodes near the *sink* send a Warning Message (WM) to nodes near the origin. After receiving WMs, nodes close to the origin send a Control Message (CM) to the sensor node. As a disadvantage, the model is not compliant with the IEEE 802.15.4 and has a significant overhead concerning a high number of message exchanges.

A priority-based method is proposed by [27] to allocate network resources, maintaining fairness between the devices' communication. Although this proposal acts centrally and does not address traffic differentiation, the BS operates an auction-driven online selection scheme to define priority access considering characteristics such as cost, precision, location, and amount of data collected.

Leão et al. [10] propose mechanisms to proportionally configure the communication structures of cluster-tree WSNs. Among them, the *proportional Superframe Duration Allocation based on the message Load* (Load-SDA) scheme defines the superframe durations and beacon intervals for clusters based on the data load generated by child nodes. Regarding load-based congestion control, the work proposed by Lino et al. [23] combines the Load-SDA scheme with a guided network formation algorithm similar to [24], providing reduced end-to-end communication delays and homogeneous branches for convergecast traffic. Also about convergence systems, Yuan et al. [25] propose an algorithm to control the monitoring load received by the base station, which can be a mobile node, aiming to obtain Quality of Service (QoS) and to save battery energy. On the other hand, considering control messages, Jing et al. [26] propose two methods for congestion control by local actuation: the first, based on the data collection that keeps a table of coordinator nodes, and the second, a local energy-based actuation, designed to schedule sleep time for the control flow, in order to overcome the limitation of control traffic in WSNs.

Within this context, it becomes clear the lack of efficient approaches to dynamically reconfigure IEEE 802.15.4 cluster-tree networks, in the presence of critical events that change the network data load. This paper aims to propose a mechanism able to dynamically reconfigure large-scale cluster-tree WSNs, in order to ensure Quality of Service for both monitoring and control traffics.

4. Problem Statement

This work assumes that sensor nodes are randomly deployed in a large-scale two-dimensional environment. These nodes are grouped into clusters according to the IEEE 802.15.4/ZigBee cluster-tree topology, and are formed using a random cluster formation process. The network may suffer from occasional load disturbances (critical events) generated by sensor nodes during the monitoring process, which may require reconfiguration of the cluster-tree parameters.

A critical event may sporadically occur in the monitoring process, implying that the data rate of one (or several) message streams must be modified. After deploying the network, each sensor node starts to collect monitoring data and establishes its default data acquisition rate. From the moment a critical phenomenon occurs in the environment, the default acquisition rate may be changed, indicating that a critical event has occurred. Thus, since new message periodicities are being imposed on the network, network overloads may occur.

In real-time IoT applications, critical events need to be reported as soon as they are detected, in order to trigger suitable protection mechanisms. In a real-world environment, temperature, humidity, pressure, and light sensors are commonly coupled to devices for large-scale control and monitoring applications. Figure 3 illustrates an example of this scenario.

This scenario involves four different types of sensor nodes: *node 1* (humidity), *node 2* (temperature), *node 3* (light) and *node 4* (pressure). In Figure 3a, message streams are highlighted to illustrate the path traveled by the data from the generator node towards the PAN coordinator. Each sensor node can be characterised in terms of the node depth, superframe duration, beacon interval, and operational load. Figure 3b shows the changes in data acquisition rates for each sensor node. For example, *node 1* identifies a change in the behaviour of its monitored physical variable, thus requiring an increase in the amount of information to be sent to the sink node.

As there is an increase in the flow of messages generated by the set of sensor nodes located within the region of the critical event, the current configuration of the cluster-tree may not be able to handle this additional load on the path along the network branches, and this may give rise to typical problems such as node overload, congestion, higher delays and packet losses. It is important to consider that since all data messages are being sent to the PAN coordinator (sink node), the problem will be more serious for CHs closer to the PAN coordinator, as they will have to deal with data accumulated from their child CHs.

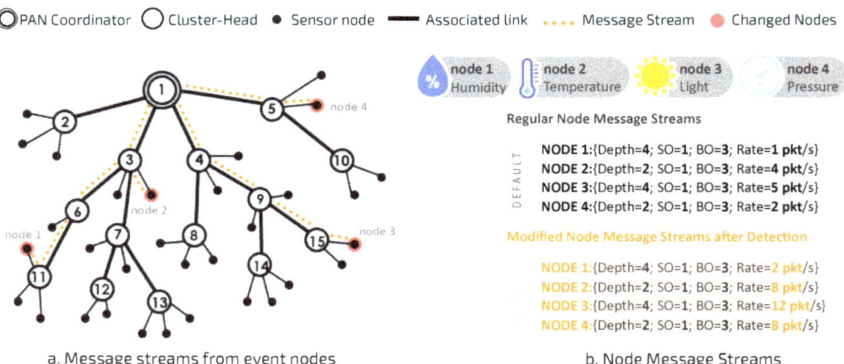

Figure 3. The critical events of nodes and the structure of messages stream generated.

Therefore, we identify a need to define efficient communication mechanisms for dynamically reconfiguring cluster-tree networks based on changes in the mode of the monitoring traffic, and the importance of performing this dynamic reconfiguration without affecting the current operation of the network.

5. Dyret: Dynamic Reconfiguration Mechanism of Cluster-Tree Wireless Sensor Networks

In this paper, we propose a new communication mechanism, called DyRET, to deal with the previous described problems. The steps of the DyRET mechanism are described in the following subsections: Section 5.1 defines the main assumptions made here; Section 5.2 describes the superframe allocation procedure; Section 5.3 presents the critical event (disturbance) detection process, that is used to notify the PAN coordinator of a requested mode change; and, finally, the reconfiguration and notification processes for clusters are explained in Sections 5.4 and 5.5.

5.1. Assumptions

Considering a cluster-tree WSN and their types of traffic, this work considers the following assumptions:

- All sensor nodes are randomly deployed and uniformly distributed throughout the environment.
- There are no mobile nodes in the environment.
- The cluster-tree network follows a random formation process, started by the PAN coordinator, which is CH of the first cluster. Other nodes are randomly selected to be cluster-heads, thus being able to form their own clusters. The cluster formation procedure is recursively performed until all sensor nodes have been associated with a CH.
- The PAN coordinator is the root and sink node of the cluster-tree network.
- The cluster scheduling is based on one-collision domain, where only one of the clusters is active at any particular time.
- Sensor nodes are not aware of their location or of any other information about their neighbourhood on the network.
- Cluster-heads do not perform any filtering, fusion or aggregating procedure.
- All monitoring traffic generated by sensor nodes is forwarded to the *sink node*, following a multi-hop tree-based routing.

- The control traffic (downstream messages), if any, simultaneously occurs with the monitoring traffic (upstream messages).
- Sensor nodes monitor events at regular intervals within a specified time frame (periodic traffic).
- Sensor nodes are able to inform the PAN coordinator that they have modified the periodicity (data rate) of a specific message stream (critical events).

Please note that although this work assumes that the cluster-tree is formed randomly, any type of cluster-tree can be dynamically reconfigured by the proposed DyRET mechanism.

5.2. Data Acquisition-Based Superframe Duration Allocation Process

Figure 4 illustrates a random network formation process and the use of a proportional superframe duration allocation procedure to initially configure the cluster-tree network.

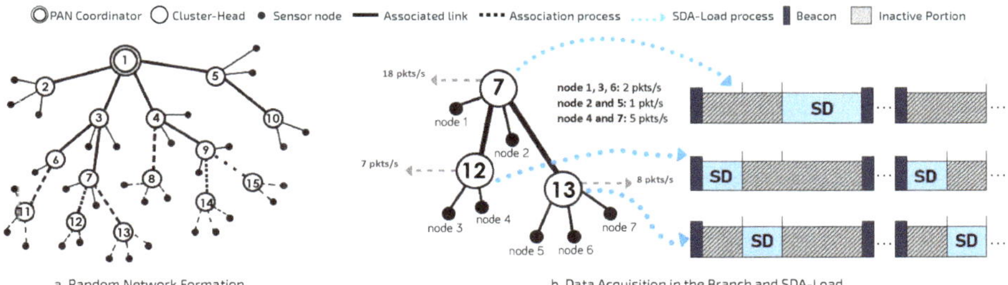

Figure 4. The cluster-tree network model: (**a**) an example of random cluster-tree formation process; (**b**) the allocation of superframe durations proportional to the data load imposed by sensor nodes (Load-SDA scheme).

After finishing the random network formation process (Figure 4a), DyRET considers that superframe durations of clusters are scheduled in order to avoid collisions between data and beacon frames, as described in [32]. Thus, each cluster has its own superframe duration defined by the Load-SDA scheme proposed in [10]. This approach defines proportional superframe durations considering the data load imposed on each CH by sensor nodes, as shown in Figure 4b. Then, as soon as all sensor nodes are associated with a specific CH, each one of them identifies its data acquisition rate, which is defined as its standard data rate and it is considered by Load-SDA scheme.

According to the Load-SDA scheme [10], each sensor with a message stream S_i periodically generates a message that is sent to the sink node (PAN coordinator) through the tree routing. Each message stream is characterised by the data message size and its generation periodicity, imposing a network use factor. In this way, the size of the beacon interval must be large enough to be able to handle all superframe durations. At the same time, BI should be as short as possible in order to reduce end-to-end communication delays. Thus, we have:

$$\sum_{j=1}^{N_{CH}} SD_j \leq BI \leq P_{min} \tag{2}$$

where SD_j is the superframe duration allocated to CH_j, BI is the beacon interval, N_{CH} is the total number of cluster-heads generated in the cluster-tree network, and P_{min} corresponds to the shortest data rate period within the set of message streams generated by the sensor nodes.

In addition, Kohvakka et al. [33] models the required time T_{TXD} to transmit a single data frame, as follows [33]:

$$T_{TXD} = T_{BACK} + T_{PKT} + T_{TX\_RADIO} + T_{ACK} \tag{3}$$

where T_{BACK} is the total *backoff* period and T_{PKT} is the packet transmission time, which denoted by $\frac{L_{PKT}}{Rad}$ (L_{PKT} corresponds to data frame size and Rad is the radio data rate). T_{TX_RADIO} corresponds to time duration the radio takes to switch between different operation modes and T_{ACK} corresponds to acknowledgements transmission time, denoted by $\frac{L_{ACK}}{Rad}$ (L_{ACK} is the acknowledgement frame size).

Considering Equation (3), Leão et al. [10] have estimated the number X of messages transferred over a minimum superframe duration SD_{min} as follows:

$$X = \left\lfloor \left(\frac{SD_{min}}{T_{TXD}} \right) \times p_s \right\rfloor, \tag{4}$$

where SD_{min} corresponds to the duration when $SO = 0$ (value only *aBaseSuperframeDuration*) and p_s is the probability of a successful transmission.

In this way, we can initially define the number of SD_{min} required for each cluster-head CH_j according to the data load imposed by sensor nodes of the branch by applying Equation (5):

$$SD_j = \left\lceil \frac{\sum_{i \in S_{below}} \left\lceil \frac{\frac{1}{P_i}}{BI} \right\rceil}{X} \right\rceil \times SD_{min} \tag{5}$$

where $\sum_{i \in S_{below}} \left\lceil \frac{\frac{1}{P_i}}{BI} \right\rceil$ corresponds to the maximum number of messages generated by the set of child nodes of the cluster-head CH_j (including the accumulated message traffic of child coordinators), with data periodicity P_i during a beacon interval (BI).

5.3. Implicit Notification Process of Critical Events Using the Data Frame Reserved Field

After defining the superframe durations for cluster-heads and starting the monitoring process, sensor nodes are responsible to identify and notify the PAN coordinator about any detected critical event. Event notification is reported between sensor nodes and PAN coordinator by using reserved bits in the data frame. The approach used in this work is known as ICN [30], where notification bits are transmitted using a *piggyback* technique in the MAC frame header—MHR (Figure 5) to identify the change of data acquisition of a particular node and to alert the PAN coordinator about a critical event.

Bits: 0-2	3	4	5	6	7-9		10-11	12-13	14-15
Frame Type	Security Enabled	Frame Pending	Ack Request	PAN ID Compression	Reserved		Dest. Addressing Mode	Frame Version	Source Addressing Mode
					X	Multiplicity			
					0/1	00/01/10/11			

Figure 5. Detail of the dataframe MHR format, modified from the IEEE 802.15.4 standard [6].

In this work, three bits are used to notify a critical event through the reserved field. The most significant bit is used to identify a specific network reconfiguration round, in order to avoid more than one reconfiguration process to be triggered for the same critical event. In turn, the two least significant bits are used by sensor nodes to represent the multiplicity of their processed data acquisition rates ("00", "01", "10" or "11"). Table 1 shows the different possible behaviours for sensor nodes used in this work.

As described in Table 1, a sensor node can operate at its default data rate, setting its bits to "00", or else it can change its acquisition rate to the double of default load (less significant bits set to "01") or four times the default load (bits "10"). In turn, a sensor node can decrease its acquisition rate by setting its bits to "11", when a critical event is finished. As previously described, the most significant bit 'X' is used to identify whether a given data packet belongs to a current reconfiguration process or if it corresponds to a modification in the data rate of a sensor node. For example, when the network is fully deployed and the monitoring is started, the default load operated by each device is set to "000" (where

X = '0' identifies the current operation and the multiplicity = "00" as the default data rate). If a set of sensor nodes identifies a new critical event and they change their default acquisition rate to twice, their bits must be changed to "001". Then, the PAN coordinator will be able to identify this mode change request and trigger a reconfiguration procedure (if needed). After the network reconfiguration is complete, sensor nodes will change their bit X to 1 (able to identify a new critical event) and reset the multiplicity value to "00".

Table 1. Table indicating notification bits and degree of network behaviour change.

Bits	Description
X00	Regular flow
X01	2 × increase of the data flow periodicity
X10	4 × increase of the data flow periodicity
X11	2 × decrease of the data flow periodicity

It is important to highlight that the proposed notification mechanism does not require any modification of the structure of the data frame, maintaining the compliance with IEEE 802.15.4 standard. Upon receiving data packets with a mode change request (modified multiplicity bits), the PAN coordinator will be able to start a new network reconfiguration procedure (if needed), which will be detailed in the following subsections.

5.4. Reconfiguration Analysis and Calculation

The PAN coordinator is responsible for performing the necessary reconfiguration calculations for the cluster-tree network, according to the received multiplicity bits from sensor nodes. The objective is to verify the need for recalculating the main communication structures of CHs (SDs and BIs), in order to avoid possible network overloads or network congestion issues. Figure 6 illustrates this situation.

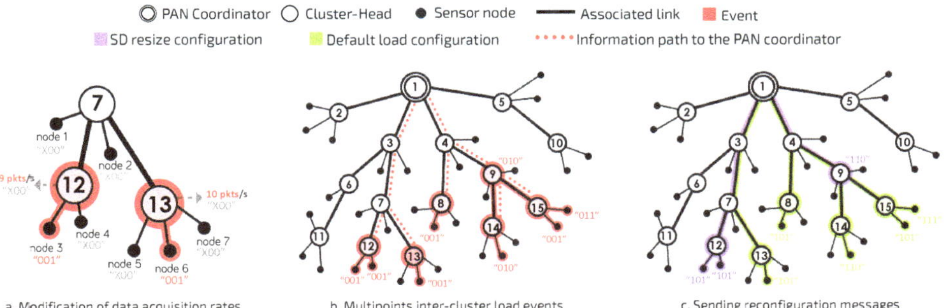

Figure 6. Network behaviour considering the details of critical event detection and reconfiguration.

Figure 6a,b illustrate the scenario where several sensor nodes can detect and report a critical event in the monitored environment. Within this context, the PAN coordinator applies the Load-SDA algorithm again, in order to recalculate the BO and SO values for each of the involved CHs, but considering the new load imposed by sensor nodes affected by the critical event. In the following, PAN coordinator must analyse the impact upon the current configuration and verify whether a new set of superframe durations is required and if it is schedulable (according to Equation (2)).

On the one hand, if the new superframe reconfiguration does not impact the current configuration (the same superframe durations allocated to all CHs), PAN coordinator only send (reset) control messages to sensor nodes with changed multiplicity bits in order to inform that, from this moment, the current data rates for each one them become their default data rates (green flow shown in Figure 6c).

On the other hand, if the new superframe reconfiguration is different of the current superframe configuration for CHs, and meets Equation (2), the PAN coordinator will send control messages for CHs, containing the new reconfiguration for SO and BO values. Moreover, PAN coordinator send (reset) control messages to sensor nodes with changed multiplicity bits, becoming their new default data rates. Notice that changing the SD for a given CH can cause the subsequent CHs to shift in the scheduling structure (as shown in Figure 7).

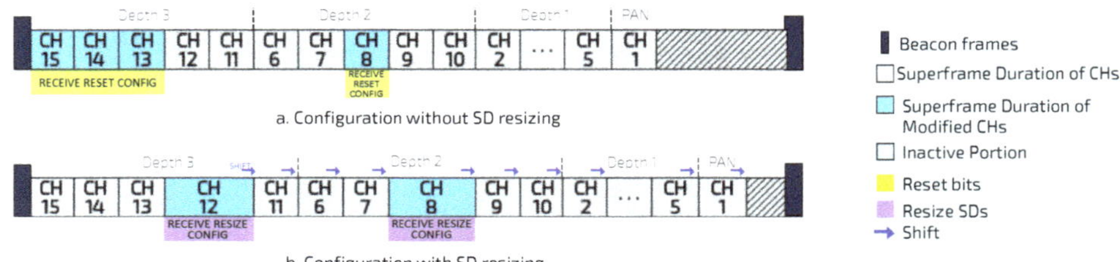

Figure 7. Communication structures after the reconfiguration analysis.

Furthermore, if the new superframe reconfiguration does not meet Equation (2), the new set of generated superframes will not be schedulable (because it does not fit within the BI, or because the required BI should be longer than the minimum period). Thus, the reconfiguration scheme proposed in this work considers that the PAN coordinator can gradually decrease the data acquisition rate (data rates) of all sensor nodes in the network (not involved in the critical event). As a consequence, the total network load is reduced until it becomes schedulable (to fit inside the BI). In this case, PAN coordinator must send control messages composed of the new values of SO and BO for CHs, in addition to the value corresponding to the reduction rate for non-event sensor nodes.

Finally, considering the superframe reconfiguration described in this subsection, the PAN coordinator is responsible for notifying all the involved nodes. For this, DyRET uses an opportunity window mechanism in order to quickly broadcast control messages (downstream traffic). This mechanism is described in the following subsection.

5.5. Opportunity Window and Dissemination of Reconfiguration Control Messages

To promote a self-adaptive system and to dynamically reconfigure communication structures, DyRET considers an Opportunity Window (OW) mechanism. OW allows the implementation of a hybrid scheduling model, that temporarily changes the current bottom-up scheduling to a top-down scheduling, in order to prioritise the control traffic. Moreover, this mechanism also promotes the fast control message dissemination through an improved configuration of the CSMA-CA parameters, as described in [34]. Figure 8 illustrates the OW mechanism for a depth-4 cluster-tree network.

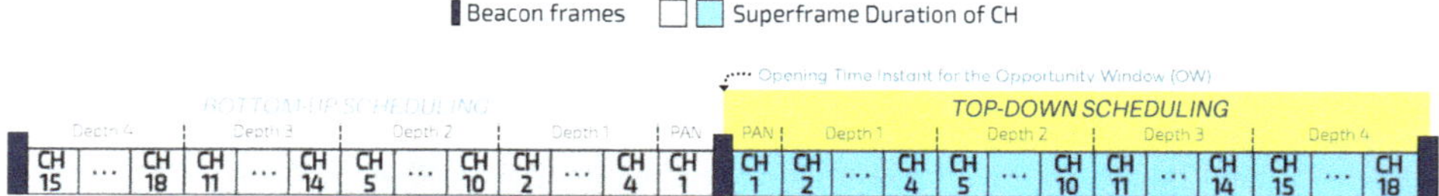

Figure 8. Rescheduling model addressed for prioritisation different traffics.

Before sending control messages with the new reconfiguration during the top-down scheduling, the PAN coordinator is responsible for creating a set of warning messages (*WARN_msg*) and forwarding them to all descendants CHs during the bottom-up scheduling. This mechanism is intended to individually notify each CH about the correct opening time instant for the Opportunity Window, avoiding thus temporal inconsistencies.

Each *WARN_msg* is composed of a tuple <#, *D*, *R*>, where # corresponds to the sequence number of the warning message, *D* is the maximum depth of the cluster-tree network and *R* corresponds to the redundancy value, representing the number of replicas of the warning message that the PAN coordinator will send. Upon receiving at least one of the warning messages *WARN_msg*, each CH can define the number of remaining BIs for the opening time instant for the OW, through Equation (6):

$$N_{BI} = (D - d_i) + (R - \#),\qquad(6)$$

where N_{BI} is the number of remaining beacon intervals for creating the OW and d_i is the depth of CH_i.

Figure 9 illustrates the timeline of creating the OW for a cluster-tree network with a maximum depth D of 4 and a redundancy value R of 3.

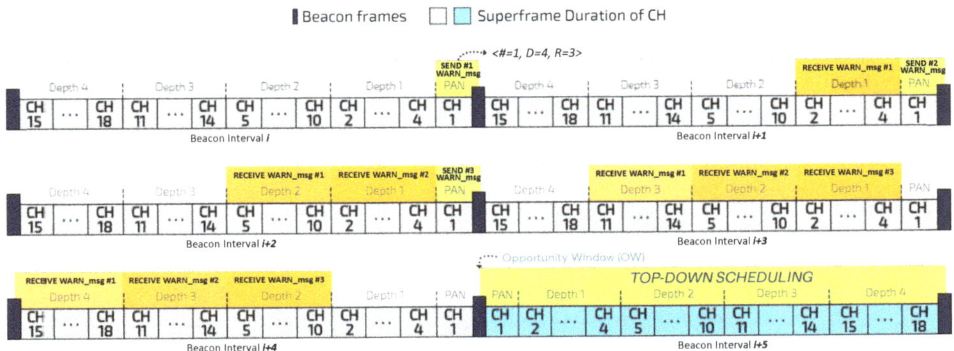

Figure 9. Timeline of the process of creating an Opportunity Window.

Importantly, warning messages are sent to CHs across the network through the indirect communication mechanism provided by the IEEE 802.15.4 standard. In indirect communication, a coordinator node indicates in the pending address field of its beacon that data is pending to be transferred. Each child node will inspect the beacon frame to verify if its address is pending. If so, this node requests the data from the coordinator during the CAP. In turn, the coordinator receives this request and subsequently sends the pending data during the CAP period, using the CSMA-CA algorithm. After receiving the data, the child node confirms its reception.

Considering the correct time instant to open the OW, each CH performs the change from bottom-up to top-down scheduling according to Equation (7):

$$TDSched_{CH_i} = 2 \times BI - 2 \times offset[CH_i] - SD[CH_i]\qquad(7)$$

where $TDSched_{CH_i}$ is the new offset for CH_i in the top-down scheduling; and the $offset[CH_i]$ and $SD[CH_i]$ are, respectively, the initial offset and the superframe duration of cluster-head CH_i.

Therefore, after the definition of the opportunity window, the PAN coordinator will start the dissemination of reconfiguration control messages throughout the network, which are forwarded to all CHs through an indirect communication mechanism. To guarantee a higher probability of accessing the wireless channel, the sending of reconfiguration control message among the coordinator nodes is carried out by changing the default values of the *macMinBE* and *macMaxBE* variables, according to the strategy proposed in [34].

After all CHs have received the reconfiguration control messages, the bottom-up scheduling is reestablished and the monitoring traffic is prioritised again, until a new critical event is identified and the entire reconfiguration process is restarted. To establish the bottom-up scheduling, each CH calculates its new beacon sending time *ReconfSched* based on received reconfiguration information through Equation (8):

$$ReconfSched_{CH_i} = offset[CH_i] + SD[CH_i] + new\_offset[CH_i]\qquad(8)$$

where $new\_offset[CH_i]$ is the new offset calculated during reconfiguration for CH_i.

Algorithm 1 describes the proposed DyRET mechanism. Please note that the PAN coordinator is responsible for performing the main steps of DyRET. Although these operations may require higher processing power and energy consumption, the PAN coordinator is commonly a special node with more memory and computational power. Furthermore, the processing time for this type of equations is negligible.

Algorithm 1: DyRET Algorithm.

---

1 /* The following code is executed for all cluster-heads   */
2 **repeat**
3   **if** (*CurrentScheduling == BottomUp*) **then**
4   **if** (*CH receives a WARN_msg <#, D, R>*) **then**
5   // CH calculates the number of BIs N_{BI} for creating the OW
6              $N_{BI} = (D - d_i) + (R - \#)$;
7   // CH calculates its new offset in the top-down scheduling
8              $TDSched_{CH_i} = 2 \times BI - 2 \times offset[CH_i] - SD[CH_i]$;

9   **else**
10   // CurrentScheduling is Top-Down
11   **if** (*CH receives a reconfiguration control message*) **then**
12   CH updates its SD and offset for the bottom-up scheduling;
13   // CH calculates the instant to reestablish the Bottom-up scheduling
14             $ReconfSched_{CH} = offset[CH] + SD[CH] + new\_offset[CH]$;

15 **until** *the cluster-tree is not operational*;

16 /* The following code is executed by the PAN Coordinator   */
17 **repeat**
18   PAN Coordinator receives the data frames generated by sensor nodes;
19   **if** (*CurrentScheduling == BottomUp*) **then**
20   **if** (*PAN Coordinator identifies a data frame with modified multiplicity bits*) **then**
21   **if** (*The need of recalculating the SDs and BI for CHs == True*) **then**
22   **repeat**
23   PAN Coordinator applies the Load-SDA based on the new load imposed by nodes;
24   PAN Coordinator recalculate the BO and SO values for involved CHs;
25   **if** (*Set of SDs is schedulable*) **then**
26   PAN Coordinator sends WARN_msg <#, D, R> for CHs;

27   **else**
28   PAN Coordinator gradually decrease the data acquisition rate of all sensor nodes;

29   **until** *the set of SDs is schedulable*;

30   **else**
31   // CurrentScheduling is Top-Down
32   PAN Coordinator sends reconfiguration control messages for all involved nodes during the OW;
            repeat
33   **if** (*All CHs received their reconfiguration messages*) **then**
34   // The Bottom-up scheduling is reestablished
35                  $ReconfSched_{CH_i} = offset[CH_i] + SD[CH_i] + new\_offset[CH_i]$;

36   **until** *all CHs have received the reconfiguration control message*;

37 **until** *the cluster-tree is not operational*;

---

6. Simulation Assessment

This section details the simulation assessment of the event-triggered dynamic reconfiguration mechanism proposed in this work. This assessment compares the behaviour of a network that uses the DyRET mechanism vs. a network that does not use dynamic network reconfiguration in the occurrence of critical events. The target of this assessment is to highlight how the DyRET communication mechanism is able to handle the efficient dissemination of both monitoring upstream messages and reconfiguration downstream control flow messages, avoiding typical cluster-tree network impairments, such as high end-to-end communication delays, network congestion, and high packet loss rates.

CT-Sim [35] has been used to assess the performance of the proposed mechanisms. CT-Sim is a set of simulation models based on *Castalia* [36], which implements the main features of cluster-tree networks.

6.1. Simulation Scenarios

For this simulation assessment, we consider three different communication scenarios (Figure 10), each one with three different number of nodes (100, 150 and 200 nodes, plus the PAN coordinator). For the sake of convenience, the following terms are used to describe the different simulation scenarios:

- *Monitoring*—Monitoring Network without events: monitoring environment without the occurrence of critical events and without the implementation of the dynamic reconfiguration mechanism;
- *Event-Only*—Monitoring Network with events and without dynamic reconfiguration: monitoring environment with the occurrence of critical events; however, no dynamic reconfiguration mechanism is implemented;
- *DyRET*—Monitoring Network with events and dynamic reconfiguration: monitoring environment implementing the proposed DyRET mechanism, to deal with the occurrence of critical events and to dynamically reconfigure the network.

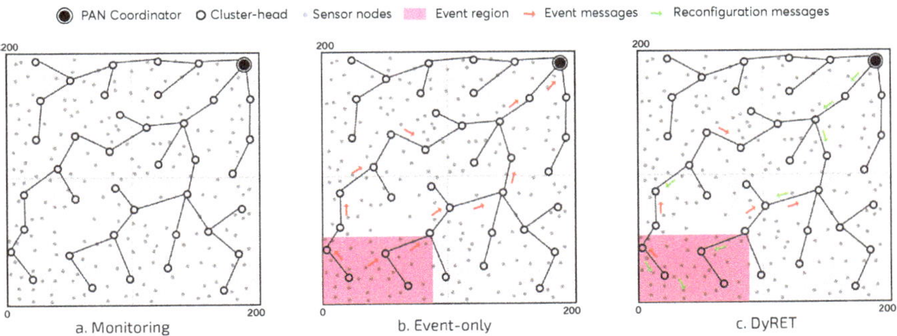

Figure 10. Different simulation approaches assessed.

The cluster-tree formation process is based on the IEEE 802.15.4 standard. The PAN coordinator is located at the corner of the environment (195 m $\times$ 195 m) and it is responsible for starting the network formation process, by associating sensor nodes, forming its own cluster and selecting a set of child nodes to be cluster-heads. Each CH (including the PAN coordinator) can associate a maximum of 6 (six) child nodes and select a maximum number of 3 (three) candidate child nodes to be CH, which can generate their own clusters.

Regarding the monitoring traffic, after the cluster-tree formation, each sensor node generates 2000 data messages at a data rate of 0.05 pkts/s (periodicity of 1 packet every 20 s), which are forwarded across the network to the PAN coordinator (sink node). Importantly, CHs do not perform any data aggregation or fusion procedure, which implies that all monitoring traffic is routed to the sink node. The superframe durations (defined by SO parameters) are proportionally allocated to each CH according to the data load of the descendant nodes (implemented by the Load-SDA algorithm [10]). In turn, BI is defined according to Equation (2). For this simulation assessment, as the shortest message periodicity P_{min} is 1 packet every 20 s, the value of BO parameter was defined to 10 (BI of 15.72 s).

For the *Monitoring* scenario (Figure 10a), no critical events are generated, i.e., the existing traffic is just the typical monitoring traffic generated by sensor nodes. This scenario is used as the basis to assess the impact of inserting critical events upon cluster-tree networks and the benefits of using a dynamic reconfiguration mechanism.

On the other hand, in the *Event-Only* and *DyRET* scenarios, there is the occurrence of critical events (Figure 10b,c). This simulation study considers the occurrence of a single critical event, where the event area is defined as a rectangular region located at the opposite corner to the PAN coordinator, comprising sensor nodes localised in the range of 80 m × 50 m (about 10% of the sensor nodes). Furthermore, sensor nodes within the critical event area have their data acquisition rates changed with multiplicity 4 (*bits* "10"), which is equivalent to modify its periodicity from 1 packet every 20 s to 1 packet every 5 s. The critical event was scheduled to take place at 1000 s of simulation. Thus, event sensor nodes will send their data messages considering this new periodicity. As the critical event remains until to finish the simulation, each sensor node will maintain its new periodicity for sending all its data messages (defined to 2000 packets).

Table 2 summarises the main configuration parameters used in the simulations. The *macMaxFrameRetries* parameter corresponds to maximum number of packet transmission retries and its value was set to 3 (default value) [6]. In this simulation assessment, we have adopted the IEEE 802.15.4-compliant CC2420 radio and the unit disc model as the propagation model. Moreover, we used an advanced wireless channel model based on empirically measured data, available in Castalia simulator [36].

Table 2. Simulation parameter configuration.

Description	Value
Environment area	200 m × 200 m
Number of sensor nodes (except PAN Coordinator)	100/150/200
Critical event area	80 m × 50 m
Critical event occurrence	1000 s
Monitoring messages (per node)	2000 packets
macMaxFrameRetries	3
Beacon Interval Size	15.72864 s
Multiplicity of events	4 × default load
Simulation Time	60,000 s
Number of seeds (per scenario)	11 rounds
Fairness interval of results	95%
Radio model	Chipcon CC2420
Radio propagation model	Unit disc model

6.2. Results and Discussion

Considering the proposed methodology and the described simulation scenarios, the following performance metrics will be considered:

- **Communication end-to-end delay:** time interval between the data frame generation at the application layer of the source node and its reception at the application layer of the destination node (sink);
- **Packet loss rate:** percentage of packets that are lost during the communication period, considering all the discarded messages due to lack of buffer space, lost messages due to collisions, and/or transmission failures;

- **Network reconfiguration time:** the number of *beacon* intervals required to send all reconfiguration control messages and thus, to reconfigure the overall network.

Firstly, the average end-to-end communication delay and average packet loss rate for all sensor nodes were assessed, considering all simulation scenarios and the aforementioned approaches. Then, we evaluate the same network metrics considering only the sensor nodes located at the region of the critical event, in order to compare the obtained results when using or not using the reconfiguration scheme. Finally, the network reconfiguration time is analysed based on the number of BIs required to send all reconfiguration control messages.

The results and discussion are presented in the following subsections.

6.2.1. Discussion of Results Considering All the Sensor Nodes of the Cluster-Tree Network

To demonstrate how critical events significantly affect the behaviour of cluster-tree networks, Figure 11 illustrates the average end-to-end communication delays for all simulation scenarios, considering all the sensor nodes of the network. It can be considered that the Monitoring approach presents the base scenario, against which any comparison should be made.

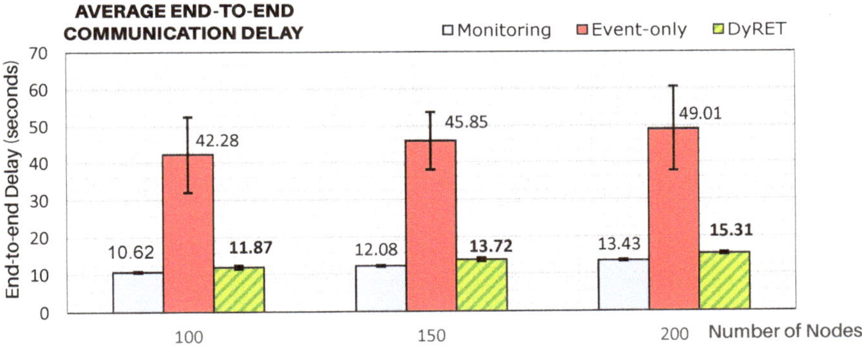

Figure 11. Average end-to-end communication delay for all sensor nodes.

The modification of the data acquisition rate of sensor nodes located at the region of the critical event can cause serious effects to the end-to-end communication delays, if no efficient action is taken. As expected, it can be observed that end-to-end delays for the Event-only approach are remarkably higher (about 4 times higher) than for the base case of just Monitoring traffic. It can also be observed the effectiveness of the proposed DyRET communication mechanism to handle the dynamic reconfiguration of a cluster-tree network. Using the DyRET mechanism, the end-to-end delays can be significantly reduced, even in the presence of critical events, keeping these results compatible with the scenario without the occurrence of critical events (Monitoring scenario).

6.2.2. Discussion of Results Considering Sensor Nodes Involved in the Critical Event

Another relevant result is to assess the network behaviour of data flows generated by sensor nodes involved in the critical event. Figure 12 illustrates the average rates of message discarded for the data flows generated by sensor nodes located at the region of the critical event, considering all simulation scenarios and all analysed approaches.

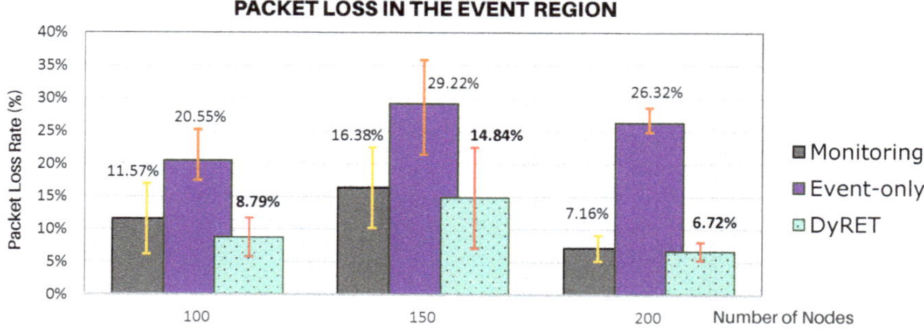

Figure 12. Average packet loss rate for sensor nodes involved in the critical event, considering all simulation scenarios.

As it can be observed in Figure 12, the Event-only approach presents a much greater number of discarded messages due to the critical event, when compared to the DyRET approach. As Event-only approach does not implement any mechanism to adequately reconfigure the communication network, the increase of data acquisition rate induces a quicker buffer occupation, which causes a larger number of discarded messages due to buffer overflows. On the other hand, as the DyRET approach reconfigures the communication network in the presence of critical events, data messages are quickly disseminated along the network, allowing alleviating the overload of the buffers and avoiding further message discards.

Moreover, Figure 13 illustrates the timeline of discarding messages for both the Event-only and DyRET approaches. It can be observed that, until the occurrence of the critical event (1000 s), the average packet loss rates present similar values for both approaches.

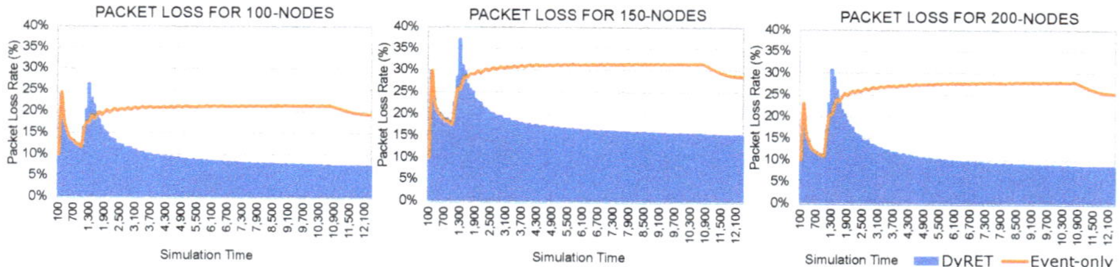

Figure 13. Timeline of packet losses (the range evaluated is 0 until the time the values remain constant).

After the occurrence of a critical event, the DyRET mechanism quickly recovers from a peak of packet loss during the actuation period (about 60 to 90 s). During this period, control messages are concurrently sent to the sensor nodes for the reconfiguration of the network. As long as the reconfiguration process is complete, the average packet loss rate is reduced until it remains constant, and at a similar value as for the Monitoring Scenario. This is not obviously the case of the Event-only approach, which linearly grows until it reaches its maximum peak.

In addition, Figure 14 illustrates the average end-to-end communication delays for sensor nodes located in the region of the critical event. It shows that DyRET approach presents significantly smaller end-to-end communications delays (close to Monitoring approach) for all simulation scenarios, even with the increase of the acquisition rate of sensor nodes at the critical event region. In turn, as Event-only approach does not implement any online reconfiguration mechanism, a higher message periodicity will cause a cumulative effect upon the buffers of cluster-heads belonging to the branch of the tree until the PAN coordinator, which will generate higher end-to-end communication delays and higher packet loss rates.

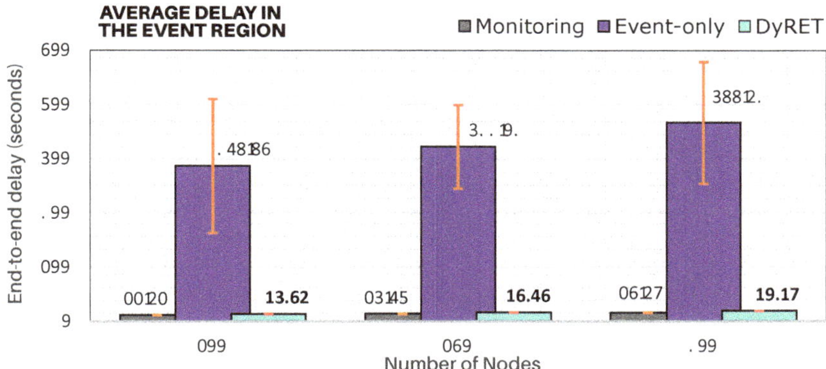

Figure 14. Average communication delay for sensor nodes involved in the critical event region.

Moreover, Figure 15 presents the timeline of the average end-to-end communication delay for both the Event-only and DyRET approaches. It is cleat that, after the occurrence of the critical event (1000 s), the average end-to-end delay highly increases for Event-only approach, while the proposed DyRET approach keeps almost constant delay rates. These results illustrate the relevance of using efficient network reconfiguration mechanisms when the behaviour of data flows is changed along the cluster-tree operation.

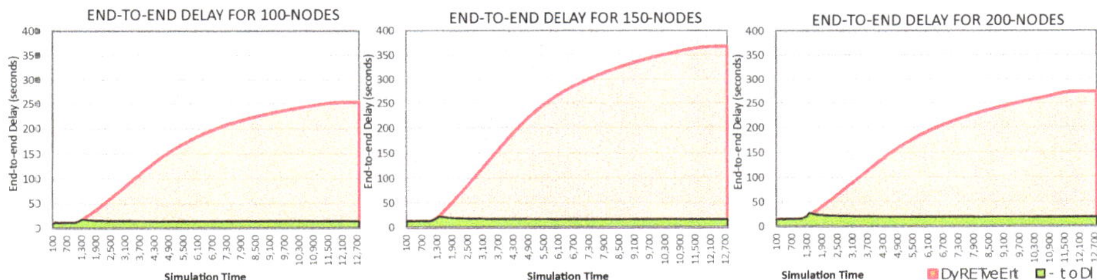

Figure 15. Timeline of delays (the range evaluated is 0 until the time the values remain constant).

Importantly, the end-to-end communication delay and packet loss rates in the 150-nodes scenario are higher than the 200-nodes scenario. As the network formation procedure is random, event nodes can be located in different branches and depths for the simulation scenarios. For the 150-nodes scenario, event-nodes were located at the deepest branches (average depth of 8), when compared to the 200-node scenario (average depth of 7).

6.2.3. Discussion of Results About the Network Reconfiguration Time

Finally, we have also assessed the time spent to reconfigure the cluster-tree network, from the instant of the occurrence of a critical event until the network is completely reconfigured.

Figure 16a illustrates the ratio between the required number of beacon intervals (OW size) with the average maximum depth of a cluster-tree WSN. Considering that a beacon interval is approximately 15 s, a network with a maximum average depth of 7 requires an Opportunity Window size of 4 BIs (approximately 1 min) for the overall network reconfiguration (for the 100-nodes and 150-nodes scenarios). For 200-nodes scenario, around 5 beacon intervals ($\approx$ 84 s) are required to send reconfiguration messages for the entire network. Such values correspond to a reconfiguration time in seconds as outlined in Figure 16b.

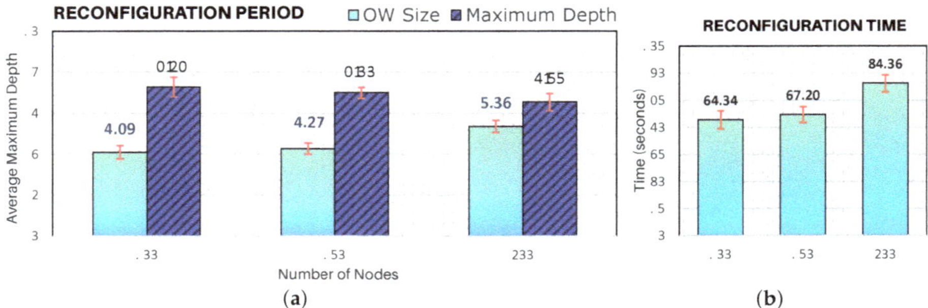

Figure 16. Reconfiguration time. (**a**) The ratio of the number of BIs in the Opportunity Window under the maximum network depth; (**b**) the respective time in seconds.

It is important to notice that despite the significant increase of the density of the communication environment, the size of the OW remains low. This fact illustrates that the configuration of CSMA-CA parameters during the Opportunity Window period combined with the hybrid scheduling actuation model is crucial for the efficient dissemination of control messages (downstream traffic).

Furthermore, the total actuation time is composed by the sum of the reconfiguration time plus the Opportunity Window configuration time. This OW configuration time comprises the period between the PAN coordinator to identify the first data packet with modified bits and the last *WARN_msg* being received by sensor nodes. Table 3 illustrates the average total actuation time for all simulation scenarios.

Table 3. The average total actuation time for all scenarios.

Scenario	Warning Period (BI)	Warning Period (s)	Actuation Time (BI)	Actuation Time (s)
100 nodes	11.58	182.16	15.53	244.36
150 nodes	12.06	189.72	16.58	260.88
200 nodes	13.51	212.61	18.88	296.97

Finally, and for the sake of clarity, Figure 17a illustrates the average occupancy rate of superframes for all simulation scenarios. The horizontal blue bars represent the sum of active periods of clusters before the occurrence of the critical event and then the red bars represent the increase in seconds caused by the critical event, representing thus the new sum of superframes of clusters after the reconfiguration process.

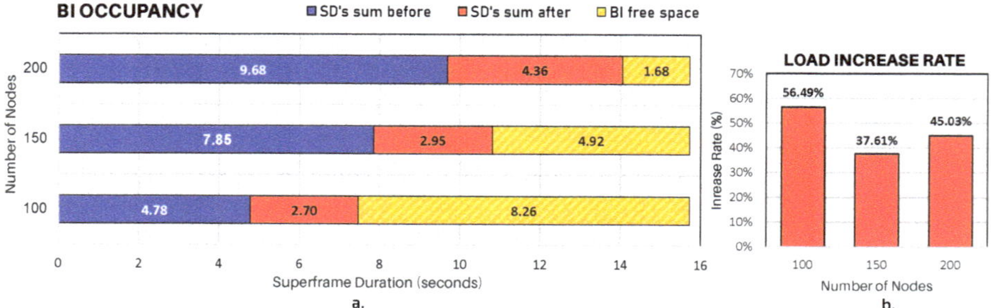

Figure 17. Ratio between load imposed and the behaviour of structures: (**a**) the load consumed and the free space within BI; (**b**) load increased percentage caused by event-nodes.

Simulation results have shown that DyRET mechanism is able to improve the transmission of data messages. It is worth mentioning that whenever a critical event is triggered, i.e., there is a disaster situation evidence, all the sensor nodes located in that region must increase their sensing data rate to send relevant information to a BS. As a consequence, the convergecast traffic is increased across all branches that form the path of this information to the BS. DyRET mechanism significantly increases the quality of service of data transmission when compared with a traditional approach, being adequate to be used in real-world disaster situations.

7. Conclusions

This paper proposes a mechanism called DyRET (*Dynamic REconfiguration of cluster-Tree wireless sensor networks*) based on the IEEE 802.15.4 standard. The communication mechanism in DyRET aims to increase the quality of service for the dynamic reconfiguration of cluster-tree networks, thus reducing end-to-end communication delays, congestion of the network and packet loss rates.

The main idea underlying DyRET is the detection of critical events that causes changes in the data acquisition rates of sensor nodes in order to allow the PAN coordinator to efficiently reconfigure the cluster-tree network without impacting the typical monitoring traffic. To achieve this, we propose a set of communication mechanisms that identify critical events and notify the PAN coordinator, reconfigure the communication structures based on critical events, and quickly disseminate the reconfiguration messages for the involved nodes, without impacting the monitoring traffic and while maintaining network synchronisation.

A simulation assessment was performed to evaluate the behaviour of the proposed DyRET mechanism in comparison to approaches that do not use a dynamic reconfiguration scheme. Through the use of implicit event notification and an opportunity window mechanism, we have shown that DyRET can reduce the packet loss rate and the end-to-end communication delays, even with increases in data rates resulting from the occurrence of critical events.

The simulation results illustrate that the DyRET scheme can reduce the end-to-end communication delays by a factor of up to 20 in environments where sensor nodes modify their data rates by an average factor of four compared to the default data load. In addition, the dissemination of control messages within the opportunity window allows all network nodes to be reconfigured within four or five beacon intervals.

A critical event occurrence may increase data rate transmission and, consequently, it triggers the construction of a new beacon scheduling by the PAN coordinator. However, there are situations where this scheduling is only feasible if some cluster-tree branches can reduce their sending rates as discussed in this paper. Then, as future work, we intend to extend DyRET by the use of mechanisms to better balance the network load, such as data fusion or aggregation, allowing parts of the network in stable situations to reduce their rates further. Moreover, we are planning to implement the DyRET mechanism in a real-world scenario testbed, for example, in a fire detection region with a high-temperature critical event. Finally, we aim to integrate the DyRET with guided cluster-tree formation procedures to obtain more-balanced cluster-tree networks.

Author Contributions: E.L. and F.V. proposed the DyRET mechanism; M.L. designed the simulation models; E.L., C.M., F.V. and M.L. proposed the assessment framework; M.L. and A.S. performed the simulation assessments and analysed the resulting data; F.V., R.M. and C.M. provided guidance for writing and revised the paper. All the authors contributed to the writing of this document. All authors have read and agreed to the published version of the manuscript

Funding: The authors would like to acknowledge the support from FAPEPI/MCT/CNPq/CT-INFRA n° 007/2018, CNPq/Brazil (Universal Project 443711/2018-6), CAPES/Brazil (PrInt CAPES-UFSC "Automação 4.0") and FCT/Portugal (Project UIDB/50022/2020) funding agencies.

Conflicts of Interest: The authors declare no conflict of interest.

Abbreviations

The following abbreviations are used in this manuscript:

BI	Beacon Interval
BO	macBeaconOrder
BS	Base Station
CAP	Contention Access Period
CFP	Contention Free Period
CH	Cluster-Head
CKN	Connected K-Neighbourhood
CM	Control Message
CSMA-CA	Carrier Sense Multiple Access - Collision Avoidance
DBR	Dynamic Bandwidth Re-allocation
DCR	Dynamic Cluster scheduling Reordering
DyRET	Dynamic REconfiguration mechanism of cluster-Tree WSNs
FACC	Fairness Aware Congestion Control
FFD	Full Function Devices
GTS	Guaranteed Time Slot
ICD	Intelligent Congestion Detection
ICN	Implicit Congestion Notification
IEEE	Institute of Electrical and Electronics Engineers
IIoT	Industrial Internet of Things
IoT	Internet of Things
IWSN	Industrial Wireless Sensor Network
Load-SDA	proportional Superframe Duration Allocation based on the message Load
LPWAN	Low-Power Wide-Area Networks
LR-WPAN	Low-Rate Wireless Personal Area Network
MAC	Medium Access Control
MHR	MAC frame header
OW	Opportunity Window
PAN	Personal Area Network
PCCP	Priority-based Congestion Control Protocol
PHY	Physical Layer
PRA	Priority-based Rate Adjustment
QoS	Quality of Service
RFD	Reduced Function Devices
SD	Superframe Duration
SO	macSuperframeOrder
TDCS	Time Division Cluster Scheduling
TDCS-PCC	Time Division Cluster Scheduling-Period Crossing Constraint
WM	Warning Message
WSN	Wireless Sensor Network

References

1. Im, J.; Park, H.; Takeuchi, W. Advances in Remote Sensing-Based Disaster Monitoring and Assessment. *Remote Sens.* **2019**, *11*, 2181. [CrossRef]
2. Aponte-Luis, J.; Gómez-Galán, J.; Gómez-Bravo, F.; Sánchez-Raya, M.; Alcina-Espigado, J.; Teixido-Rovira, P. An Efficient Wireless Sensor Network for Industrial Monitoring and Control. *Sensors* **2018**, *18*, 182. [CrossRef] [PubMed]
3. Ramesh, M.V. Wireless sensor network for disaster monitoring. In *Wireless Sensor Networks: Application-Centric Design*; IntechOpen: London, UK, 2010; p. 51.
4. Hoeller, A.; Souza, R.D.; López, O.L.A.; Alves, H.; de Noronha Neto, M.; Brante, G. Analysis and Performance Optimization of LoRa Networks With Time and Antenna Diversity. *IEEE Access* **2018**, *6*, 32820–32829. [CrossRef]

5. Zuniga, J.; Ponsard, B. *SIGFOX System Description*; IETF: Fremont, CA, USA, 2016.
6. IEEE 802.15.4. IEEE Standard for Low-Rate Wireless Networks. *IEEE Std 802.15.4-2015 (Revision of IEEE Std 802.15.4-2011)*; IEEE: New York, NY, USA, 2015; p. 709. [CrossRef]
7. ZigBee. ZigBee Specification. *ZigBee Alliance (Document 053474r20)*; ZigBee Alliance: San Ramon, CA, USA, 2012.
8. Li, C.; Zhang, H.; Hao, B.; Li, J. A Survey on Routing Protocols for Large-Scale Wireless Sensor Networks. *Sensors* **2011**, *11*, 3498–3526. [CrossRef] [PubMed]
9. Erdelj, M.; Mitton, N.; Natalizio, E. Applications of Industrial Wireless Sensor Networks. In *Industrial Wireless Sensor Networks: Applications, Protocols, and Standards*; Güngör, V.Ç., Hancke, G.P., Eds.; CRC Press: Boca Raton, FL, USA, 2013.
10. Leão, E.; Montez, C.; Moraes, R.; Portugal, P.; Vasques, F. Superframe Duration Allocation Schemes to Improve the Throughput of Cluster-Tree Wireless Sensor Networks. *Sensors* **2017**, *17*, 249. [CrossRef] [PubMed]
11. Severino, R.; Pereira, N.; Tovar, E. Dynamic Cluster Scheduling for Cluster-tree WSNs. *SpringerPlus* **2014**, *3*, 493. [CrossRef] [PubMed]
12. Candell, R.; Kashef, M.; Liu, Y.; Lee, K.B.; Foufou, S. Industrial Wireless Systems Guidelines: Practical Considerations and Deployment Life Cycle. *IEEE Ind. Electron. Mag.* **2018**, *12*, 6–17. [CrossRef]
13. Vakaloudis, A.; O'Leary, C. A framework for rapid integration of IoT Systems with industrial environments. In Proceedings of the 2019 IEEE 5th World Forum on Internet of Things (WF-IoT), Limerick, Ireland, 15–18 April 2019; pp. 601–605. [CrossRef]
14. Koubaa, A.; Cunha, A.; Alves, M.; Tovar, E. TDBS: A time division beacon scheduling mechanism for ZigBee cluster-tree wireless sensor networks. *Real-Time Syst.* **2008**, *40*, 321–354. [CrossRef]
15. Al-Ghamdi, B.; Ayaida, M.; Fouchal, H. Scheduling approaches for wireless sensor networks. In Proceedings of the 2015 15th International Conference on Innovations for Community Services (I4CS), Nuremberg, Germany, 8–10 July 2015; pp. 1–6. [CrossRef]
16. Kim, S.; Kim, J. Dynamic Self-Reconfiguration Algorithms for Wireless Sensor Networks. In Proceedings of the 2010 10th IEEE/IPSJ International Symposium on Applications and the Internet, Seoul, Korea, 19–23 July 2010; pp. 91–95. [CrossRef]
17. Choudhury, N.; Matam, R.; Mukherjee, M.; Lloret, J. A Non-Threshold-Based Cluster-Head Rotation Scheme for IEEE 802.15.4 Cluster-Tree Networks. In Proceedings of the 2018 IEEE Global Communications Conference (GLOBECOM), Abu Dhabi, UAE, 9–13 December 2018; pp. 1–6. [CrossRef]
18. Thanh Hiep, P. Spatial Reuse Superframe for High Throughput Cluster-Based WBAN with CSMA/CA. *Adhoc Sens. Wirel. Netw.* **2016**, *31*, 69–87.
19. Heinzelman, W.R.; Chandrakasan, A.; Balakrishnan, H. Energy-efficient communication protocol for wireless microsensor networks. In Proceedings of the 33rd Annual Hawaii International Conference on System Sciences, Maui, HI, USA, 7 January 2000; Volume 2, p. 10. [CrossRef]
20. Yin, X.; Zhou, X.; Huang, R.; Fang, Y.; Li, S. A Fairness-Aware Congestion Control Scheme in Wireless Sensor Networks. *IEEE Trans. Veh. Technol.* **2009**, *58*, 5225–5234. [CrossRef]
21. Hanzalek, Z.; Jurcik, P. Energy Efficient Scheduling for Cluster-Tree Wireless Sensor Networks with Time-Bounded Data Flows: Application to IEEE 802.15.4/ZigBee. *IEEE Trans. Ind. Inform.* **2010**, *6*, 438–450. [CrossRef]
22. Ahmad, A.; Hanzálek, Z. An Energy Efficient Schedule for IEEE 802.15.4/ZigBee Cluster Tree WSN with Multiple Collision Domains and Period Crossing Constraint. *IEEE Trans. Ind. Inform.* **2018**, *14*, 12–23. [CrossRef]
23. Lino, M.; Vasconcelos, V.; Ian, A.; Leão, E.; Soares, A.; Montez, C. An Efficient Mechanism to Improve Convergecast Traffic in Cluster-tree Wireless Sensor Networks Based on IEEE 802.15.4. In Proceedings of the IECON 2019—45th Annual Conference of the IEEE Industrial Electronics Society, Lisbon, Portugal, 14–17 October 2019; Volume 1, pp. 2811–2816. [CrossRef]
24. Andrade, A.T.C.; Siedersberger, D.; Montez, C.; Moraes, R.; Leão, E.; Vasques, F. Data-Based Cluster-Tree Formation Scheme for Large-Scale Wireless Sensor Networks. In Proceedings of the 2018 IEEE 16th International Conference on Industrial Informatics (INDIN), Porto, Portugal, 18–20 July 2018; pp. 175–180. [CrossRef]

25. Yuan, Z.; Ouyang, Y.; Shan, L.; Hu, H.; Li, Z. A Load Balancing Algorithm in Convergent Wireless Sensor and Cellular Networks. In Proceedings of the 2012 8th International Conference on Wireless Communications, Networking and Mobile Computing, Shanghai, China, 21–23 September 2012; pp. 1–6. [CrossRef]

26. Jing, G.; Jia, L.; Wang, X.; Xian, H.; Zhang, L.; Gong, S. Reconfiguration during data collection for many-to-one routing in wireless sensor networks. In Proceedings of the 2017 Chinese Automation Congress (CAC), Jinan, China, 20–22 October 2017; pp. 5279–5282. [CrossRef]

27. Trevathan, J.; Hamilton, L.; Read, W. Allocating Sensor Network Resources Using an Auction-Based Protocol. *J. Theor. Appl. Electron. Commer. Res.* **2016**, *11*, 41–63. [CrossRef]

28. Mukherjee, M.; Shu, L.; Hu, L.; Hancke, G.P.; Zhu, C. Sleep Scheduling in Industrial Wireless Sensor Networks for Toxic Gas Monitoring. *IEEE Wirel. Commun.* **2017**, *24*, 106–112. [CrossRef]

29. Biabani, M.; Fotouhi, H.; Yazdani, N. An Energy-Efficient Evolutionary Clustering Technique for Disaster Management in IoT Networks. *Sensors* **2020**, *20*, 2647. [CrossRef] [PubMed]

30. Wang, C.; Li, B.; Sohraby, K.; Daneshmand, M.; Hu, Y. Upstream congestion control in wireless sensor networks through cross-layer optimization. *IEEE J. Sel. Areas Commun.* **2007**, *25*, 786–795. [CrossRef]

31. Zhang, J.; Yang, T. Event Driven Self-Adaptive Routing Algorithm in Wireless Sensor Network. In Proceedings of the 2013 Fourth International Conference on Emerging Intelligent Data and Web Technologies, Xi'an, China, 9–11 September 2013; pp. 323–328. [CrossRef]

32. Koubaa, A.; Cunha, A.; Alves, M. A Time Division Beacon Scheduling Mechanism for IEEE 802.15.4/Zigbee Cluster-Tree Wireless Sensor Networks. In Proceedings of the 19th Euromicro Conference on Real-Time Systems (ECRTS'07), Pisa, Italy, 4–6 July 2007; pp. 125–135. [CrossRef]

33. Kohvakka, M.; Kuorilehto, M.; Hännikäinen, M.; Hämäläinen, T.D. Performance Analysis of IEEE 802.15.4 and ZigBee for Large-Scale Wireless Sensor Network Applications. In Proceedings of the PE-WASUN '06 3rd ACM International Workshop on Performance Evaluation of Wireless Ad Hoc, Sensor and Ubiquitous Networks, Torremolinos, Spain, 6 October 2006; Association for Computing Machinery: New York, NY, USA, 2006; p. 48–57. [CrossRef]

34. Constante, L.; Lau, J.; Moraes, R.; Araujo, G.; Montez, C.; Leão, E. Enhanced association mechanism for IEEE 802.15.4 networks. In Proceedings of the 2017 22nd IEEE International Conference on Emerging Technologies and Factory Automation (ETFA), Limassol, Cyprus, 12–15 September 2017; pp. 1–8. [CrossRef]

35. Leão, E.; Moraes, R.; Montez, C.; Portugal, P.; Vasques, F. CT-SIM: A simulation model for wide-scale cluster-tree networks based on the IEEE 802.15.4 and ZigBee standards. *Int. J. Distrib. Sens. Netw.* **2017**, *13*, 1–17. [CrossRef]

36. Tselishchev, Y.; Boulis, A.; Libman, L. Experiences and Lessons from Implementing a Wireless Sensor Network MAC Protocol in the Castalia Simulator. In Proceedings of the 2010 IEEE Wireless Communication and Networking Conference, Sydney, NSW, Australia, 18–21 April 2010; pp. 1–6. [CrossRef]

Article

ISA 100.11a Networked Control System Based on Link Stability

Heitor Florencio [1,*]**, Adrião Dória Neto** [2] **and Daniel Martins** [2]

[1] Digital Metropolis Institute, Federal University of Rio Grande do Norte, Natal,
Rio Grande do Norte 59078-900, Brazil
[2] Department of Computer Engineering and Automation, Federal University of Rio Grande do Norte, Natal,
Rio Grande do Norte 59078-900, Brazil; adriao@dca.ufrn.br (A.D.N.); danlartin@gmail.com (D.M.)
* Correspondence: heitorm@imd.ufrn.br

Received: 30 July 2020; Accepted: 19 August 2020; Published: 21 September 2020

Abstract: Wireless networked control systems (WNCSs) must ensure that control systems are stable, robust and capable of minimizing the effects of disturbances. Due to the need for a stable and secure WNCS, critical wireless network variables must be taken into account in the design. As wireless networks are composed of several links, factors that indicate the performances of these links can be used to evaluate the communication system in the WNCS. This work presents a wireless network control system composed of ISA 100.11a sensors, a network manager, a controller and a wired actuator. The system controls the liquid level in the tank of the coupled tank system. In order to assess the influence of the sensor link failure on the control loop, the controller calculates the link stability and chooses an alternative link in case of instability in the current link. Preliminary tests of WNCS performance were performed to determine the minimum stability value of the link that generates an error in the control loop. Finally, the tests of the control system based on link stability obtained excellent results. Even with disturbances in the network links, the control system error remained below the threshold.

Keywords: industrial wireless sensor networks; ISA 100.11a; wireless networked control systems; link stability

---

1. Introduction

The implementation of new communication technologies in industrial automation allows for the integration of industrial processes with greater efficiency, availability and quality. The systematic association of control and monitoring systems with communication systems generates several benefits. Wireless networked control systems acquired significant technological advancements with the rise of wireless networks, advanced control, embedded computing and cloud computing.

The connection of devices spatially distributed for different purposes with wireless networks provided a great increase in applications using wireless sensor networks (WSNs). However, critical process control constraints defined some limits of WSN technology. Industrial wireless sensor networks (IWSNs) are a specific field of WSNs, which takes into account reliability constraints, timing deadlines and critical nature of industrial applications.

The industrial wireless sensor networks are being used in various branches of industry: area monitoring, structural monitoring, disaster prevention and control systems. Many applications in different monitoring and control systems are said to be safety related. In such kinds of systems, a system failure may put people in danger, lead to environmental damages or result in economic losses [1]. The guarantee on packet reception and reliability must also be provided for feedback control systems to operate properly. There is also a need for extensive measures to be implemented to counter the uncertainties of wireless means of communication [2]. System performance evaluation parameters are required to ensure these extensive measures.

The development of wireless networked control systems (WNCSs) is fundamental in Industry 4.0. The development of both intelligent manufacturing equipment and intelligent control systems is a priority in the machine tools sector. Additionally, the priorities in the area of IT include the Internet of Things (IoT) and its applications, including industrial control [3]. Since the cyber-physical systems represent the integration of physical systems with computing and networking capabilities, WNCSs are an important class of cyber-physical systems in Industry 4.0, in which physical processes are controlled using wireless sensors, actuators and controllers [4].

A WNCS connects sensors and actuators of a plant to a controller via a wireless network, which has several critical communication channels. These links are classified as critical because they are part of several closed-loop control systems. Thus, the system design must also take into account an evaluation parameter for these links. Link stability is an appropriate factor for analyzing links that carry information from sensors to the controller.

In this paper, we present an ISA 100.11a wireless network control system. The control system receives two level measurement values, but the choice of the level value depends on the stability of these two links. Link stability assesses the performance of the link from samples of the received signal strength. A permanent monitoring of the link stability guarantees the regularity of the control system. The networked control system is implemented with ISA 100.11a wireless sensors, network manager, gateway, controller and wired actuactor (pump).

The remainder of this paper is organized as follows: Section 2 discusses some related works about WNCS and link stability on wireless networks. In Section 3, the overall the system architecture is described. Section 4 describes the software implemented in the controller of the system. All system tests and results are presented in Section 5. Finally, conclusions are stated in Section 6.

2. Related Works

2.1. Wireless Networked Control Systems

The industrial networks are part of the structure of automation systems. They allow the communication of all instruments in the system, including sensors, actuators, controllers and data acquisition stations. The wireless sensor networks have been used in many monitoring applications for various physical phenomena, such as temperature, flow, level, vibration, humidity and pump analysis [5]. With the advent of the Industrial Internet of Things (IIoT), the availability of fast, secure and reliable communication networks deployed within factories and connecting all the elements of industrial control systems became a requirement [6]. However, deployment of wireless communication in the control systems creates new obstacles to overcome, such as the need of designing the control parameters associated with the network parameters.

Wireless networked control systems allow all or some of the measurement and control signals to be transmitted over wireless channels. There are approaches in which both signals the one from sensors and the one sent to the actuators travel through wireless communication technology. However, there are other approaches that use wireless and wired signals. The data that flow between sensor nodes and controllers are not necessarily symmetric in WNCSs [7–9].

The objectives of the networked control system are to ensure that the closed-loop system has desirable dynamic and steady state response characteristics, and that it is able to efficiently attenuate disturbances and handle network delays and loss [7]. The main communication problems are the delay and the packet loss rate, which directly influence the reliability of the system. The delay problem may greatly reduce the performance of the control system, so that the stability of the system is narrow.

There are several industrial wireless communication protocols that allow different configurations of parameters and structures. Likewise, there are several control loop techniques with several variations. A proper model must consider the parameters of both control and communication. Efficient integration of communication and control has been identified as a high-impact challenge for the next generation of industrial automation systems [10]. A more integrated approach is necessary in order to design systems that systematically model parameters between the communication and control systems.

Determining the optimal parameters for minimum network cost while achieving feasibility is not trivial because of the complex interdependence of the control and communication systems. WNCSs require novel design mechanisms to address the interaction between control and wireless systems for maximum overall system performance and efficiency [7]. Several researches have being developed to model, evaluate and validate wireless network control systems [11].

Park et al. [7] presented and explain many criticial system variables. There are critical variables both in the control system and in the wireless communication system, which are closely linked. For instance, the control system defines the sampling period and the communication protocols determine the retransmission mechanism in case of failure. Therefore, the maximum retransmission period (communication system) must be determined together with sampling period (control system). The critical variables in the communication aspect are packet delay rate and packet loss rate. Additionally, in the control system aspect, the variables are sampling period, message delay and message dropout [7].

The delay time is a parameter used by several researchers to model WNCSs with different approaches [5,10,12]. Shi et al. [5] modeled the network time delay in the multi hop network caused by the S-MAC communication protocol. The model also considers the controller model in a WNCS. Arauo et al. [10] took into account the delay in the sensors' and actuators' links and modeled a solution to compensate for the delay. The paper categorizes two types of delays: delays in the access to the communication channel and delays due to the transmissions and computation at the controller. In [12], the model is based on research area of delay-constrained wireless communication. This paper analyzed a WNCS with multiple control systems sharing a commom wireless channel.

Many researchers have implemented WNCSs with simulators only, which causes the absence of experimental tests with instruments of the manufacturers in the market. By way of example, Horvath et al. [4] presented a simulation framework which includes a realistic model of the physical layer with multi-channel frequency-hopping mesh networks. The simulation framework is evaluated by implementing a WNCS based on WirelessHART.

Park et al. [7] and Araujo et al. [10] presented experimental tests with wireless instruments in the level control system of coupled tanks. Both instruments used in the tests are Telos nodes [13]. Ahlén et al. [14] presented an implementation on an industrial process at the Iggesund paper mill. All control loops were implanted using wireless sensors and actuators. Additionally, the ABB AC800M controller received and sent information to the instruments by communicating with the network gateway via Profinet. The results indicate that it is feasible to use wireless control for continuous production. Ahlén et al. [14] ensured that it is possible to reach the desired availability with wireless instrumentation compared with the wired instrumentation.

In this study, we implemented a wireless networked control system based on ISA 100.11a instruments from the manufacturer Yokogawa Electric. In addition to the delay time, the packet loss rate in communication links is also an essential parameter in the evaluation of the communication system. As such, this study evaluated the implementation of a WNCS based on the evaluation of the links that are part of the control loop. The stability of the links is the factor used in the system proposed in this work.

2.2. Link Stability in IWSN

A wireless sensor network contains several instruments located in different locations, which generate different communication links. Communication between two devices can have more than one link However, each link has its characteristics and can be affected differently by interference. One way to evaluate these links is by the link stability factor. The link stability should not be confused with the stability of the control system, the concepts are significantly different from one another.

Some researchers claim that link stability indicates how stable the link is and how long it can support communications between two nodes [15]. For others, the link stability means the link will sustain for a long time and does not break regularly [16]. Overall the link stability indicates the level of variation of the link with respect to the level of noise and the rate of packets lost.

In paper [17], we presented a study about link stability in IWSN. Link stability is defined by the variation of received signal strength (RSS) and packet delivery rate (PDR). An unstable link presents a high variation of signal attenuation and low packet delivery rate. A stable link has no signal attenuation variation and has a high packet delivery rate [17]. Table 1 is formulated to present the selected papers in [17] and their parameters used to define stability.

Table 1. Papers that define link stability and its parameters.

Paper	Year	Distance between Nodes	Link Expiration Time	Packet Delivery Rate	Received Signal Strength	Sensor Networks
Zou and Tao [18]	2012	✓	N	N	N	N
Sun et al. [19]	2010	✓	N	N	N	N
Boukerche et al. [20]	2017	✓	N	N	N	N
De Rango and Francesca [21]	2012	✓	N	N	N	N
Zhang et al. [22]	2012	✓	N	N	✓	N
Sharma and Pathak [23]	2015	N	✓	N	N	N
Idoudi et al. [24]	2016	N	✓	N	N	N
Chama et al. [25]	2013	N	✓	N	N	N
Al-Qhdah et al. [26]	2016	N	✓	N	N	N
Wang et al. [27]	2010	N	✓	N	N	N
Li and Yan [28]	2017	N	N	✓	✓	N
Priya et al. [29]	2017	N	N	✓	N	✓
Wenqing [30]	2011	N	N	N	✓	N
Brahmbhatt et al. [16]	2015	N	N	N	✓	N
Patil and Patil [31]	2015	N	N	N	✓	N
Model proposed: Florencio and Neto [17]	**2019**	N	N	✓	✓	✓

✓ : Covered	N: Not Covered

Many papers use the distance between nodes and expiration time because they are applied to mobile networks, which have a high level of mobility. Models that use link expiration time do not consider the instability during the period that the links are active on the network. The appearance of several interferences is common, even with the device in the same position in IWSN.

In industrial wireless sensor networks, the devices are typically distributed at fixed locations without mobility, and thus the links remain active throughout the period of operation. Therefore, the model of this work does not consider the link expiration time. Additionally, the distance between nodes parameter is used indirectly in the acquisition of the received signal strength parameter.

The link stability function is based on the variation of the received signal strength. It was necessary to create a factor that indicates the degradation of the signal to calculate this variation. This factor is generated from previous samples of the received signal strength and the current value. The entire process of generating link stability is detailed in Section 4.1. A link with low stability directly implies the performance of the control system.

In this study, the link stability was be used in the controller program to choose the measurement value used in the control technique. However, the evaluation of the implementation of the network control system is the main objective of this work. The next sections detail the system implemented.

3. System Architecture Design

The system architecture consists of a controller to receive data from the sensors and send the signal to the actuator. Data are collected through the ISA 100.11a network gateway, which interfaces with all measurement elements. However, the only actuation element in the system uses a wired signal. The systems in papers [4,32] also use the wired signal from the actuator.

The system implements a tank level control loop with wireless instrumentation. The system configuration allows the controller to choose the process variable (PV) value between the LD01 sensor route and the LD02 sensor route. Thus, the stability metric of these links is the parameter that defines the choice of PV. Next the controller runs the program and sends the wired signal to the pump.

The controller provides two communication interfaces with the ISA 100.11a gateway to collect data from the links and measurement variables of the instruments. In addition, an interface with a supervisory system is provided to enable monitoring of variables. Figure 1 shows the system architecture.

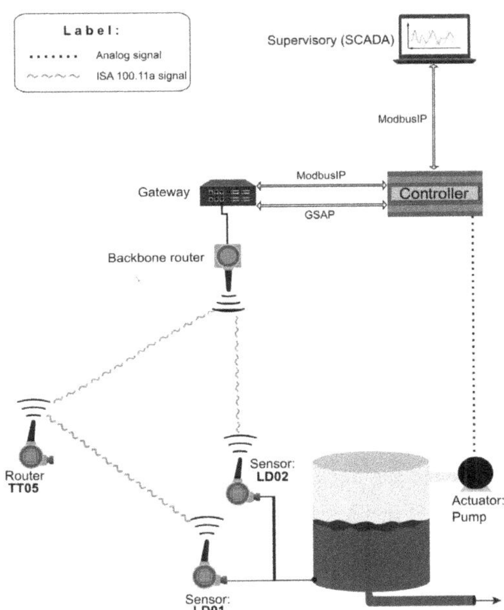

Figure 1. ISA 100.11a network control system architecture.

It is possible to notice in Figure 1 that there are two differential pressure sensors to measure the tank level (process variable): LD01 and LD02. These sensors measure the same differential pressure value. That means the controller can receive the same process value from both routes to control the tank level.

3.1. Level Control System

The control system is composed of two tanks coupled, a water basin, two level sensors and a pump. The liquid in the lower tank flows to the water basin and a pump is responsible for pumping water from the basin to the upper tank. The liquid in the upper tank flows to the lower tank. There is a sensor in a each tank for measuring the level of tank.

The system in this project used only an upper tank. The tank shown in Figure 1 represents the upper tank of the double-tank coupled. However, the two sensors LD01 and LD02 were placed at the same measuring point. The integration of these measurements from the tank with the controller was performed through the ISA 100.11a wireless network.

3.2. ISA 100.11a Network

Nowadays, industrial wireless networks are part of the structure of automation systems. They allow the communication of all field instruments, including sensors and some actuators, with controllers and data acquisition stations. Some protocols define the rules and techniques for wireless communication at the sensors, actuators and controller, such as IEC 62734 (ISA 100.11a) [33].

The ISA 100.11a architectures contain elements that perform information transmission, information routing and network management at different levels. Each network is formed with nodes and composed of a processing unit, a radio, memory, a data acquisition board and a battery. Currently, regardless of topology, all networks need a central element to concentrate information from all nodes: the network manager. The manager receives and sends the data packets to the nodes, and manages all links formed in the network. Within end-to-end communication, at least one of the elements is the manager, either by receiving the packet from a measuring element or by sending the packet to an actuation element.

The network manager is not responsible for the junction authorization of a new device to the network. The tasks of managing security keys, such as authenticating, generating and storing, are the responsibilities of the security manager. Physically, the two managers are integral logical parts of the gateway element, as shown in Figure 2a. In this work, the network manager, security manager and gateway were implemented in a single management station: YFGW410 (Yokogawa Electric) [34]. The first device in Figure 2b is gateway YFGW410.

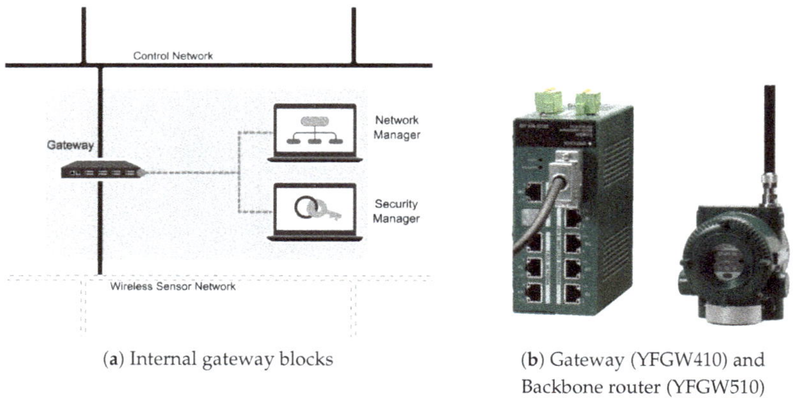

(**a**) Internal gateway blocks   (**b**) Gateway (YFGW410) and Backbone router (YFGW510)

Figure 2. Gateway, network manager, security manager and backbone router.

The second device in Figure 2b is the backbone router YFGW510 (Yokogawa Electric) [35]. The wireless devices communicate with the gateway and therefore with the managers through backbone router, which operates as an access point of wireless network.

The field devices on the ISA 100.11a network can be routers or non-routers. The use of routers in networks can transmit their data and their neighbors' data to the manager, thereby increasing the redundancy and availability of the network due to alternative paths generated.

The ISA 100.11a network of the Figure 1 architecture includes only three instruments: two differential pressure sensors (LD01 and LD02) and a temperature transmitter, which operates in the router mode of the LD01 instrument. Unlike the LD01 sensor, the LD02 instrument has a direct link to the gateway. Figure 3 shows the number and nomenclature of the system links.

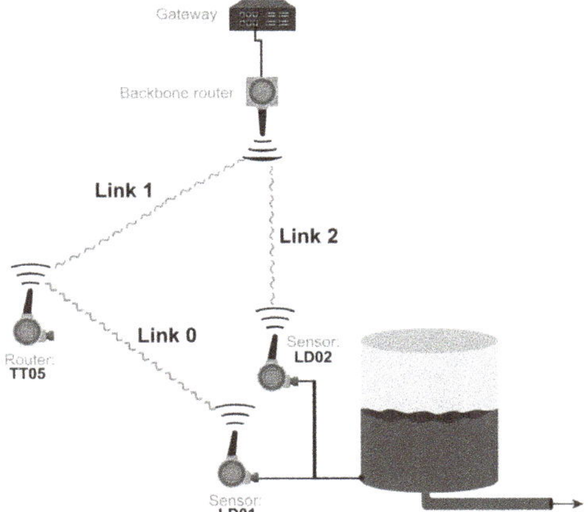

Figure 3. The system links.

The links created in the system configuration are:

- textbfLink0: LD01 → TT05.
- textbfLink1: TT05 → Gateway.
- textbfLink2: LD02 → Gateway.

The LD01 sensor route is composed of link0 and link1, and the LD02 sensor route is composed only of link2. The purpose of insertion of the router TT05 in this work was to be able to cause a forced attenuation in the antenna of this instrument and then analyze the influence on the level control loop. Therefore, it is mandatory that the controller be able to collect the link stability information before executing the control logic and sending the signal to the pump.

3.3. Controller Board

The controller must be able to collect the measurement values of the sensors (LD01 and LD02), calculate the link stability levels, execute the level control program, control the pump and send the monitoring data to the supervision system.

The link stability metric is generated from the RSSI data of the links. Hence, the controller will collect these data from the links through a communication interface with the gateway. The controller must send requests to the gateway following the GSAP (gateway service access point) specification. The system collects the measurement values from the sensors through a Modbus TCP communication interface with the gateway. Finally, the sending of data to the supervision system also uses a Modbus TCP interface.

Due to the high communication and processing capacity and the need for a WiFi module to implement Modbus TCP and GSAP commands, the system controller chosen was the ESP32 microcontroller from the company Espressif. Figure 4 shows the ESP32 controller.

Figure 4. ESP32 controller.

ESP32 is a single chip designed with ultra low power technology that incorporates microprocessing, memory, peripherals and communication modules (WiFi and Bluetooth). The main features that distinguish it from other platforms used in embedded systems are: two processing cores, a 160 MHz clock, an integrated Bluetooth module, a flash memory expandable to 32 GB, 36 GPIO pins and 18 channels of analog-to-digital converters [36].

This controller has been used in several IoT applications due to its processing power and low power consumption. It is possible to define a completely wireless solution using the ESP32 module, which integrates the IEEE 802.11 network protocol with the IoT architecture [37,38].

Two communication modules between controller and gateway were implemented: Modbus TCP communication and GSAP communication. The Modbus TCP interface is responsible for acquiring the process variable data from the LD01 and LD02 sensors. The GSAP module requests the information from the links, focusing on the RSSI values used in the link stability function. Figure 5 shows the modules implemented in the ESP32 controller.

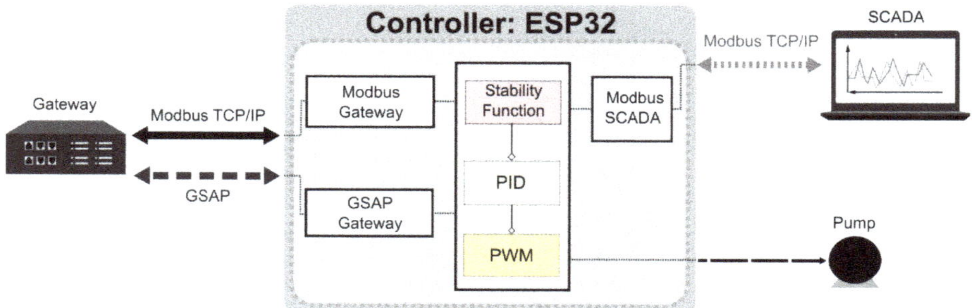

Figure 5. Controller modules.

The link stability module implements the link stability analytical model and defines which PV value will be used in the control logic. Finally, the pump receives a pulse-width modulation (PWM) signal resulting from the program that controls the tank level.

3.3.1. Modbus TCP Communication

The Modbus protocol is an industrial communication protocol at the application layer that follows a master–slave topology in order to perform the communication between devices. Only one device, the master, can initiate request–response messages to other devices (slaves) by sending a query to an individual slave or sending a broadcast query to all slaves. In the case of Modbus TCP/IP, the slave address is identified by an IP address [39].

It is then possible to use Modbus over serial protocols (e.g., RS-485) or TCP/IP protocols on Ethernet. In any case, the message structure is always the same. The Internet community can access Modbus at a reserved system port 502 on the TCP/IP stack. Modbus TCP/IP is mostly used in the data sharing between the field device level (e.g., PLC, CAN J1939 to the Modbus Gateway) and the SCADA system level. Modbus TCP/IP as a protocol could support communication between field devices via TCP, i.e., between sensors, actuators and PLCs [39,40].

There are two Modbus TCP communication modules in the controller: communication with the ISA 100.11a gateway and communication with the supervision system (ScadaBR).

In the Modbus TCP communication module with the gateway, the communication master is the ESP32 controller, which makes the requests, and the communication slave is the gateway. The slave's Modbus memory mapping (gateway) contains the PV values of the two sensors, as shown in Table 2.

Table 2. ISA 100.11a gateway Modbus memory mapping.

TAG	Modbus Data Type	Offset	Description
PV1	Holding registers Holding registers	13 14	LD01 Sensor Process Variable
PV2	Holding registers Holding registers	34 35	LD02 Sensor Process Variable

The controller also provides a Modbus communication with the ScadaBR supervisory application to supervise the control system [41]. Only the main variables of the control loop and the network are monitored.

Unlike the other Modbus module, the ESP controller is the communication slave in the interface with the ScadaBR application (the communication master). The mapping of the ESP32 Modbus memory, as shown in Table 3, only contains variables of the type holding registers. Some variables occupy more than one memory position (offset) due to the representation as a float data.

Table 3. Controller Modbus memory mapping.

TAG	Data Type Modbus	Description
ModePID	Holding registers	Mode to enable setpoint change
PV1	Holding registers	LD01 sensor process variable
SP	Holding registers	Setpoint
MV	Holding registers	Manipulated variable (PID output)
PV2	Holding registers	LD02 sensor process variable
$RSSI_\alpha$	Holding registers	RSSI do link α
$DPDUTx_\alpha$	Holding registers	Number of packages delivered by the link α
$DPDUTxFail_\alpha$	Holding registers	Number of packages lost by the link α

ModePID, PV1, SP, MV and PV2 are variables of the control loop and the other variables represent the behavior of the network links. All of these variables that describe the network's performance were collected from the gateway through GSAP communication.

3.3.2. GSAP Communication

The ISA 100.11a standard describes an access point interface to gateway services: GSAP (gateway service access point). This service is generic and should be used as a common interface above the application layer of the protocol [33].

GSAP is a specification that defines the support features that allow a communication interface between the ISA 100.11a network and an external network. The standard describes how to implement messages from the GSAP specification using the 15 objects and services described in the standard. There is no complete detail, but a reference to help understand the commands [33].

The commands are implemented using objects from the protocol application layer. Each service accesses a specific type of network manager object. However, there are several commands that can manipulate these objects. Table 4 presents some GSAP services described in the ISA 100.11a standard.

In this work, only the commands G_Session_request and G_Neighbor_Health_Report request were implemented. The first service (Session) performs the opening of the GSAP session with the gateway. The second service (Neighbor_Health Report) is responsible for requesting data from neighbors on a field device on the network.

Each service has specific fields in the request and confirmation messages. The programmer must understand all fields in the package to be able to communicate with a gateway. In addition, it is only possible to request a *Neighbor_Health Report* service if the session is already open.

In order to exemplify a GSAP service request, Table 5 presents the fields of the command *Neighbor_Health_request* with the respective example values.

Table 4. GSAP services.

Service	Command
Session	*G_Session request* *G_Session confirm*
Topology_Report	*G_Topology_Report request* *G_Topology_Report confirm*
Device_List_Report	*G_Device_List_Report request* *G_Device_List_Report confirm*
Device_Health_Report	*G_Device_Health_Report request* *G_Device_Health_Report confirm*
Neighbor_Health_Report	*G_Neighbor_Health_Report request* *G_Neighbor_Health_Report confirm*
Network_Health_Report	*G_Network_Health_Report request* *G_Network_Health_Report confirm*

Table 5. GSAP command fields: *Neighbor_Health request*.

Field	Size (Bytes)	Description	Example Value
Version	1	Gateway version	OxFO
Service type	1	Command code	0x07
Session ID	4	Session identifier. Value generated after the session creation	–
Transaction ID	4	Each communication in a session generates a different ID	–
Data size	4	Size of the object to request data	–
Header CRC	4	Header check code. Header: version, type, session ID, transaction ID and data size	–
Network address	16	Network address of the instrument that you want to collect information from neighbors	–
Data CRC	4	Network address field check code	–

The command *Neighbor_Health_request* returns an object of type *NeighborHealthReport*, which stores all values about the neighbors of a given instrument on the network. The object returns a vector of elements of type *NeighborHealth*: *neighborHealthList[]*. The *NeighborHealth* structure stores all the data from the link between the transmitter, identified by the Network address field in Table 5, and the neighbor (receiver), identified by the *networkAddress* of the *NeighborHealth* structure.

The GSAP driver sends the opening session command and the command *Neighbor_Health request* for each link in the system Figure 1, thereby obtaining all data from the network links.

4. Software Implementation

The software includes all the modules shown in Figure 5. The stability function, and PID and PWMmodules are part of the main logic of the controller. The implementation of the link stability metric generation uses the values of the received signal strength indicator (RSSI). The stability values of the links are used in the selection of PV. Finally, the controller performs the PID control technique.

4.1. Method of Link Stability

The method provides an evaluation of the link stability of a IWSN. As mentioned in Section 2.2, in the context of an environment susceptible to different types of interference, the link stability is essential to evaluate network performance.

The lower the stability of the link, the higher the variation of the attenuation in the reception signal, and consequently, the greater the instability in the delivery of packets [17].

The method is based on a linear function proportional to the variation of the received signal strength (RSS) and the packet delivery rate (PDR) within a set of samples. This metric is generated from previous samples of RSS and current value of RSS and PDR. Equation (1) presents the link stability factor.

$$LinkStability = PDR * (1 - 0.04^{MME_{RSSRatio}})\tag{1}$$

The variable $MME_{RSSRatio}$ is the exponential moving average of the standard deviation values of the attenuation ratio (*RSSRatio*). The purpose of using the moving average is to generate a filter to reduce the influence of outliers, and consequently, present a factor with greater smoothness.

The moving average is based on a set of samples of the variable ($\sigma_{RSSRatio}$); it has a size that varies when a new sample appears. The amount of samples in a data window can be defined by the user when implementing the method.

The first variable to generate the Equation (1) is the rate of attenuation of the received signal: *RSSRatio*. This variable relates the current RSS value to the previous values. The purpose is to relate the current strength value to the signal attenuation in the last samples. The Equation (2) shows the *RSSRatio*.

$$RSSRatio = \frac{RSS_i}{RSS_{max}}\tag{2}$$

The variable RSS_i represents the current value of RSS and the RSS_{max} represents the maximum value of the RSS sample set. The greater the variation of *RSSRatio*, the greater the instability of the received signal strength. Thus, the value used to calculate the moving average is the standard deviation of the RSSRatio values.

As shown in Equation (1), the exponential function with fixed base and exponent MME is used to smooth the factor that multiplies the PDR. This smoothing avoids the generation of very low values of stability when the MME variable becomes very low. Figure 6 shows the flowchart for the entire method.

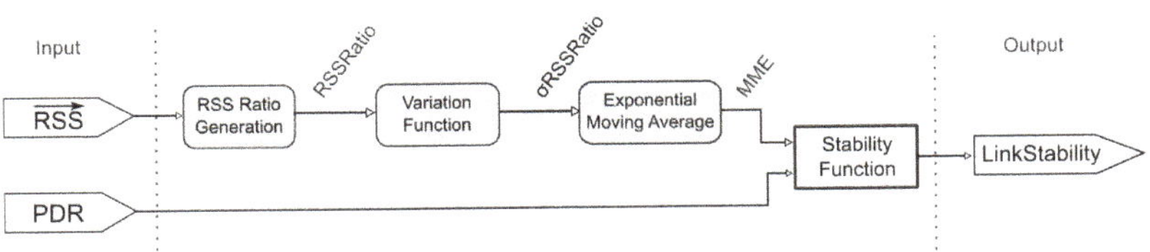

Figure 6. Overview of the method for link stability evaluation.

Part of this method was detailed and published by Florencio and Neto [17]. The link stability metric is able to detect instabilities in the links, which can cause an increase in the packet loss rate, and consequently, considering a networked control system, take the system to an unstable region [17].

4.2. Control Program

The control technique is an proportional–integral control (PI), which operates at a sampling rate of 100 ms and with gains of KP = 1.6 and KI = 0.15. As the controller modeling and tuning processes are not part of this work, standard values of other works developed with the same tank system were used.

Link0, shown in Figure 3, is the primary link to acquire the level value (PV1). However, if the stability level of link0 is equal to or below a threshold, the controller must select the PV value of the alternative route: link2 (PV2).

The control program must perform the steps listed below.

1. Read the value of PV1 (module: Modbus TCP communication with Gateway);

2. Read the value of PV2 (module: Modbus TCP communication with Gateway);
3. Read the SP value (module: Modbus TCP communication with SCADABr);
4. Read link data (module: GSAP communication with the gateway);
5. Calculate the stability value of link0 (module: stability function);
6. Calculate the stability value of link2 (module: stability function);
7. Select the PV value from the stability values (LinkStability0 and LinkStability2);
8. Execute PI control (module: PI Control);
9. Send the MV value (PWM signal) to the pump;
10. Update data for the supervisory system (module: Modbus TCP with ScadaBR).

The pseudocode Algorithm 1 presents an overview of the lines of code implemented in the ESP32 controller for level control based on the link stability metric.

Algorithm 1 Algorithm of the controller program.

```
1:  PV1 = modbusGW_request(1, 13, 2);
2:  PV1 = (double) PV1;
3:  PV2 = modbusGW_request(1, 34, 2);
4:  PV2 = (double) PV2;
5:  SP = (int) modbus_scada.Hreg(HREG_SP);
6:  gsap_requestLinks();
7:  stability0 = func_stability0();
8:  stability2 = func_stability2();
9:  if stability0 > thresholdStab then
10:     PV = PV1;
11: else
12:     PV = PV2;
13: end if
14: MV = pidTank.Compute(SP, PV);
15: PWMvalue = (int) MV;
16: analogWrite(PUMP, PWMvalue);
17: update_scadabr();
```

Each line or group of lines of code performs a step from the control program. The modbusGW_request(x, y, z) commands in lines 1 and 3 request the PV variables for each sensor, where x is the slave ID (master–slave communication), y is the address of the variable in the slave's Modbus memory mapping and z is the size of the variable (multiples of 16 bits). Hence, as the communication slave is the gateway, line 1 requests the PV value, mapped in position 13, with a size of 32 bits.

Unlike the communication of the ESP32 controller with the gateway, the controller is the slave in the communication with the between ScadaBR and controller. The modbus_scada.Hreg(x) command reads the value contained in the x position of ESP32 Modbus memory. This memory location is written in the supervision application (ScadaBR), or rather, by the operator in the Modbus writing commands.

Line of code 6 collects the network link data used to generate the link stability factor. This command updates the following code variables: RSSI0, RSSI1, RSSI2, DPDUTx0, DPDUTxFail0, DPDUTx1, DPDUTxFail1, DPDUTx2 and DPDUTxFail2. Some of these variables are used by the stability function in lines 7 and 8.

After generating the stability values, code lines 9 to 13 ensure that the LD01 (PV1) sensor value will only be used by the PI control if the stability is greater than the minimum stability threshold. Otherwise, the value of the LD02 sensor will be used by the PI control.

Finally, ESP32 performs PI control on code line 14 and sends the signal to the pump (code line 16). At the end, all data from the control loop and the network are updated in Modbus memory in order to transfer to the ScadaBR supervisory.

5. System Performance Evaluation

5.1. Implementation

A system has been developed to evaluate the performance of the wireless networked control system with ISA 100.11a devices. All elements of the architecture (Figure 1) are present in the system, as described below.

- Controller: ESP32;
- Gateway: YFGW410—Yokogawa Electric Corporation (Figure 2b);
- Access point or backbone router: YFGW510—Yokogawa Electric Corporation (Figure 2b);
- Router TT-05: YTA510 (temperature transmitter)—Yokogawa Electric Corporation;
- Level Sensor: EJX110B (differential pressure transmitter)—Yokogawa Electric Corporation;
- Actuator: Water Pump 12V;

The tank and level sensors were placed on a test bench in the Industrial Network Laboratory, as shown in Figure 7a. The backbone router and the gateway were located at a distance of 3 m from the level sensors within the same laboratory. The TT-05 router is the only instrument that was distant, in the external area, in order to perform the signal attenuation tests.

The instruments highlighted in Figure 7b are the backbone router and the gateway.

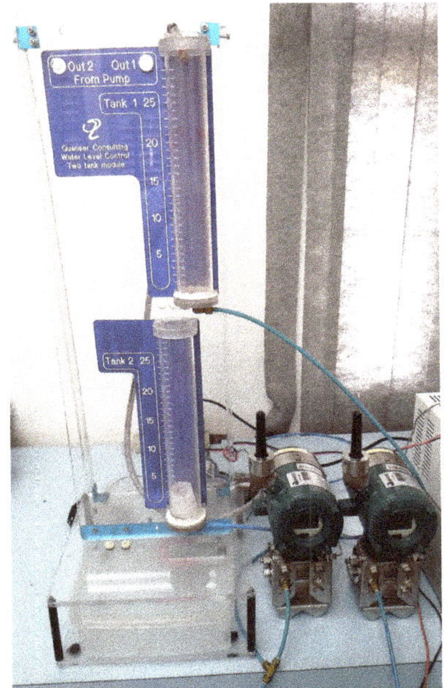

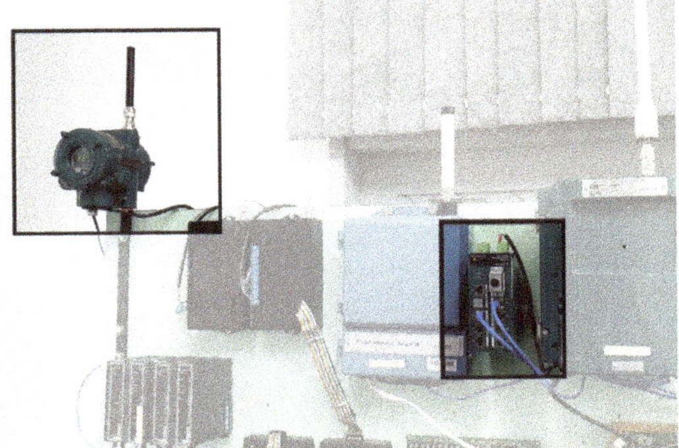

(**a**) Tank and Level Sensors   (**b**) Backbone Router and Gateway

Figure 7. System in the industrial network laboratory.

5.2. Wireless Network Level Control System: Preliminary Tests

Tests of the level control system without considering the link stability metric were performed in order to analyze the influence of stability on the control error with the system in steady state.

The methodology used in these first tests is described below.

1. Start ISA 100.11a network gateway, backbone router and instruments.
2. Launch the SCADA application.

3. Start the controller.
4. Determine the setpoint at 100 mm.
5. Execute the PI control with the PV value of the LD01 sensor.
6. Wait for the control to reach the steady state with 1% error.
7. Cause a change in attenuation in the signal of link0.
8. Collect data and analyze the influence of link failure on the controller.

The main objective of these tests is to cause a variation in the attenuation of the transmission signal of the LD01 sensor to verify its influence on the steady state of the control loop.

Step 7 of the methodology is performed only after the system reaches a steady state, considering a error of 1%. Thus, it is possible to infer that the errors that arise in the control loop are due to failures in the network link.

Seven tests were carried out with a minimum duration of 20 min, considering the time to start the controller, determine the set point value and wait for the system to reach the Steady State.

In order to present a better overview of the data in this paper, data from three tests are presented: test 01, test 02 and test 03.

5.2.1. Test 01

The first graph of test 01 shows the variables of the control loop: error (difference between the process variable and the setpoint value) and MV (manipulated variable). The controller calculated the value of the output of the PI control (variable MV), in a range from 0 to 255, to send a pulse width modulation (PWM) signal to the system actuator: the pump. Figure 8 shows the data for these variables during the test.

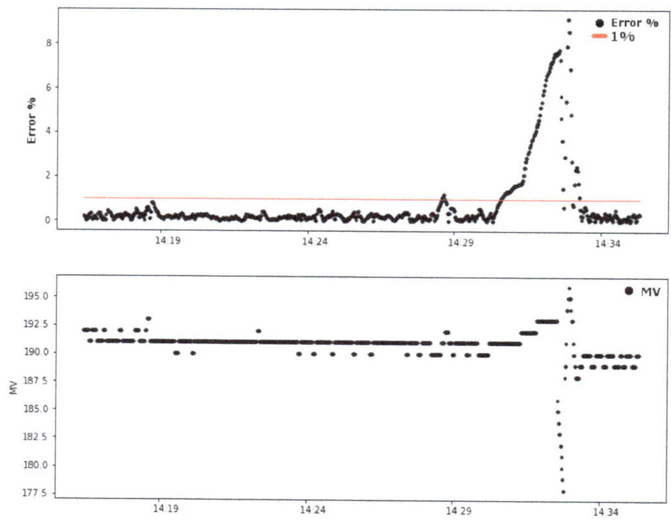

Figure 8. Test 01: Variables of the control loop (error and MV).

It is possible to observe that approximately in the time period between 14 h 30 min and 14 h 33 min the absolute error in the steady state of the control system exceeded the limits of 1%, reaching a value close to 9% of error. This change in the error naturally caused a change in the pump signal (MV), shown in the second graph of Figure 8. The red line indicates the limit of 1% error.

An attenuation on the link is forced by changing position and inserting structures that degrade the signal transmitted on the link. Figure 9 shows the relation between the link stability value and the controller error. The analysis of the influence of stability on error is discussed in this section.

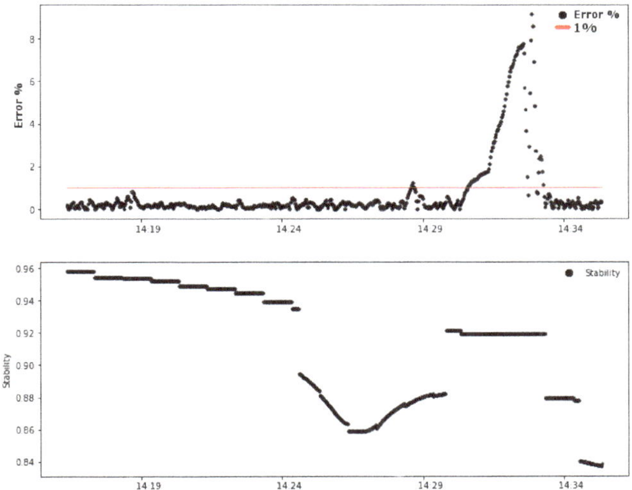

Figure 9. Test 01: Error (%) and Link0 stability.

It can be observed from Figure 9 that the variation of the link stability occurred before the variation of the control error. The vertical red dashed line indicates the moment when the forced attenuation of the link started. From these graphs, the previous influence of the link stability factor on the system error was analyzed.

5.2.2. Test 02

Test 02 also shows a change in the controller after the link attenuation. It is important to remember that the control was already in the steady state. The values of the absolute error also exceeded the limit of 1% defined as a system requirement, as shown in Figure 10.

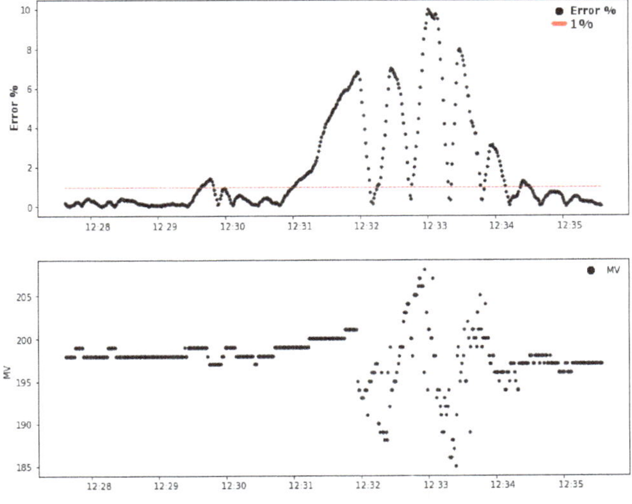

Figure 10. Test 02: variables of the control loop (error and MV).

The same behavior of test 01 happened in the second test. There is a variation in the link stability, shown in the first graph in Figure 10. Additionally, then, the control error increased in the period of time following reducing the link stability value. These behaviors of these variables are presented in the graphics of Figure 11. The dashed vertical lines in the graphs indicate the moment of the beginning of the forced attenuation.

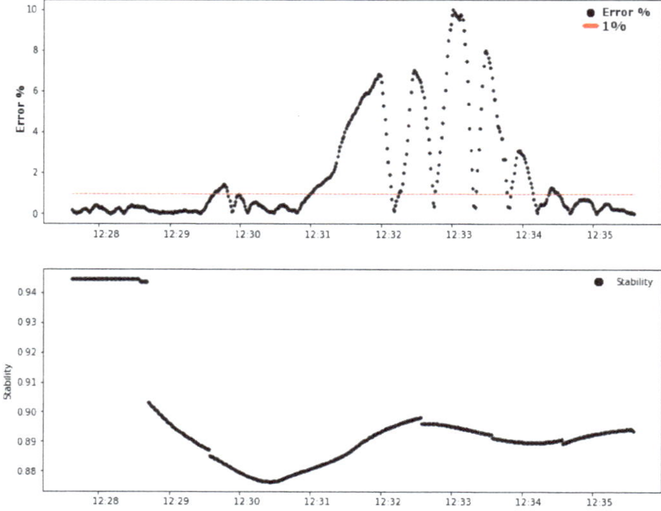

Figure 11. Test 02: error (%) and Link0 stability.

An essential step in this analysis is the verification of the rate of instantly packets delivered ratio (PDRi). The relation between the PDRi of the link and the control error is shown in the graphics of Figure 12. The variation of the control error occurred later with the increase in drops packets.

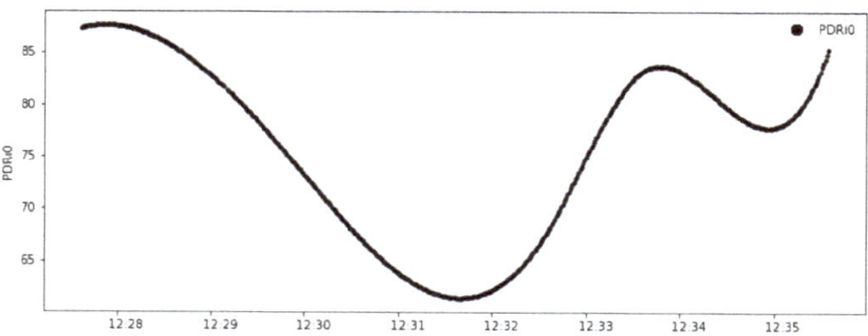

Figure 12. Test 02: Link0 PDRi.

The data presented in Figures 11 and 12 prove that the reduction of the link stability caused a variation in the packets delivered ratio, which, consequently, increased the absolute error of the control system.

5.2.3. Test 03

The results obtained in test 03 support the same conclusion as the other tests, as shown in Figure 13. The link stability factor provides a prognosis or even a prediction of the behavior of the control system in the next few minutes.

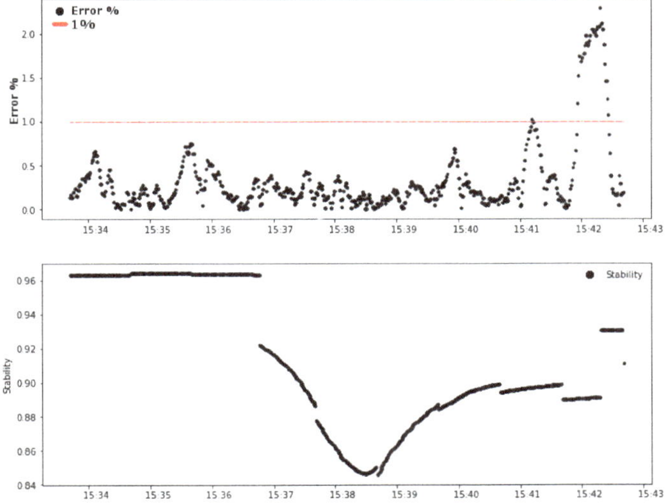

Figure 13. Test 03: error (%) and Link0 stability.

5.2.4. Preliminary Test Results

From an examination of the preliminary tests it becomes apparent that the link stability factor allows a prediction of the change in the control system error.

The second graphic in Figure 13 (test 03) shows that, in approximately 15 h 36 min 40 s, the link stability decreased to around 0.92 (92%) and remained decreasing until about 0.85 (85%). In approximately 15 h 41 min 40 s, 5 min after the decrease of the stability level, the error started its ascension until it got close to 3% error.

The values of reducing the stability factor and the time between the stability variation and the increase in error are essential to use the link stability as a decisive factor in the control loop. Thus, a summary of the data from the tests performed is presented in Table 6.

Table 6. Preliminary test results.

Test	Time to Instability	Reduction Value Range (Link Stability)
01	5m	90% a 86% (Median: 88%)
02	3m	92% a 88% (Median: 90%)
03	5m	92% a 85% (Median: 89%)
04	7m	92% a 87% (Median: 89%)
05	7m	88% a 86% (Median: 87%)
06	4m	89% a 86% (Median: 88%)
07	3m	92% a 90% (Median: 91%)

The third column of the Table 6 contains the values of the median of the reduction curve of the link stability factor. Thus, the median of these values was calculated. The median of the stability values during the reduction is 89%.

5.3. Wireless Network Level Control System Based on Link Stability

In the preliminary tests performed, the controller received the PV value from the LD01 sensor and executed the control logic. However, failure periods were observed after a reduction in the link stability value to an average of 89%. Thus, this value of 0.89 will be the threshold of link stability in control program: variable thresholdStab in Algorithm 1.

Unlike the previous results and following the architecture of Figure 1, in this final test, the controller selected the PV value based on the stability of the links. The ESP32 controller received the RSSI values of the links, stored the values and calculated the stability values of these links in real time to detect whether a variable change was required (PV1 or PV2).

Figure 14 shows that the error (%) of the controller decreased after a period of time until the end of the test.

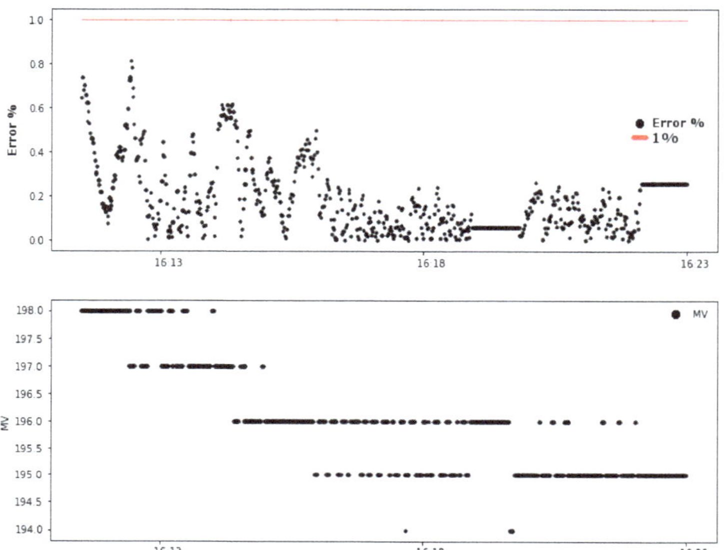

Figure 14. Result of the control system based on link stability: error (%) and MV.

The result shows that the permanence of the error below the maximum value, with low variation, ensures that the manipulated variable of the control loop remains regulated.

In this last test, the same attenuation procedures were performed for the link of the level measurement sensor. The error remained below the maximum limit due to the implementation of control logic based on link stability.

By determining the link stability values equal to or less than 89%, the controller changes the choice of the PV value. This behavior can be observed in Figure 15.

The dashed vertical red line in Figure 15 indicates the time the controller detected the link stability value less than 89% and changed the PV value to the value of the second link (Link2). This change kept the error below the threshold.

Finally, the ISA 100.11a network control system was implemented and the link stability metric was able to identify possible instabilities and prevent the failure of the system's control loop.

There are few works that performed experiments with networked control systems using WirelessHART and ISA 100.11a protocols. In addition, there is no work that implemented a WNCS based on the link stability parameter. Thus, a comparison between researches was not possible.

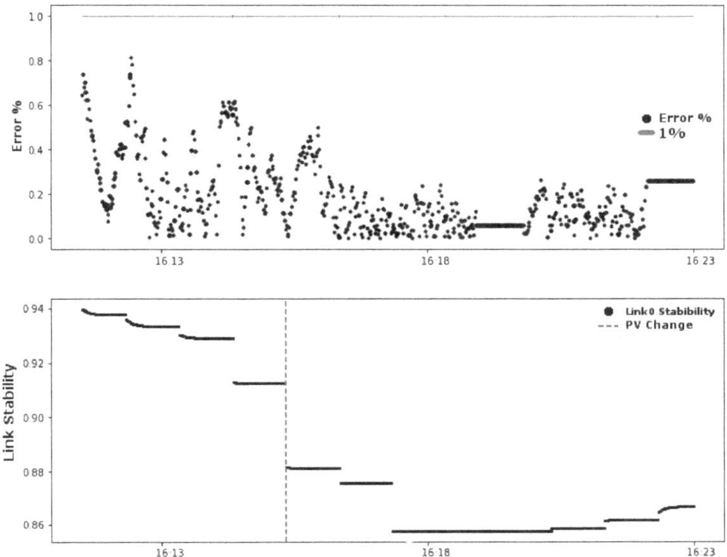

Figure 15. Results of the control system based on link stability: error (%) and link stability.

6. Conclusions

In this paper, we present the implementation of an ISA 100.11a networked control system. ISA 100.11a networks are widely used in monitoring applications. However, control applications require careful attention when designing communication and control systems. Thus, the first contribution of this work was the evaluation of networked control systems using the ISA 100.11a protocol.

The system controller uses link stability as a decisive factor in choosing PV values. The link stability model is able to detect instabilities in the communication between the instruments, and consequently, to predict failures in the control loop. Some preliminary tests were performed to analyze the behavior of the control system from the generation of purposeful noise in the system. Purposeful noises reduced the value of the link stability and then increased the error of the control loop. Thus, the proposed WCNS is based on the link stability to avoid failure in the control system. When the controller took into account the link stability, the system tests showed satisfactory results. The controller detected a low stability of the sensor link and changed the PV value to another link. The detection of link instability kept the control loop within the desirable limits. The second contribution was the use of the link stability model in a wireless network control system.

The tests were performed with ISA 100.11a instruments from the manufacturer Yokogawa Eletric. Additionally, the monitoring of the network and control system variables was done from the interface with Modbus TCP and GSAP commands. Thus, the third main contribution of this work was to provide experimental tests with instruments from manufacturers in the market.

Author Contributions: Conceptualization, H.F., A.D.N. and D.M.; Data curation, H.F. and A.D.N.; Methodology, H.F., A.D.N. and D.M.; Software, H.F.; Writing—original draft, H.F.; Writing—review & editing, H.F., A.D.N. and D.M All authors have read and agreed to the published version of the manuscript.

Funding: This research received no external funding.

Conflicts of Interest: The authors declare no conflict of interest.

References

1. Jesus, T.C.; Portugal, P.; Vasques, F.; Costa, D.G. Automated methodology for dependability evaluation of wireless visual sensor networks. *Sensors* **2018**, *18*, 2629. [CrossRef] [PubMed]
2. Raza, M.; Aslam, N.; Le-Minh, H.; Hussain, S.; Cao, Y.; Khan, N.M. A critical analysis of research potential, challenges, and future directives in industrial wireless sensor networks. *IEEE Commun. Surv. Tutor.* **2017**, *20*, 39–95. [CrossRef]
3. Kagermann, H.; Helbig, J.; Hellinger, A.; Wahlster, W. *Recommendations for Implementing the Strategic Initiative Industrie 4.0: Final Report of the Industrie 4.0 Working Group*; Forschungsunion: Berlin, Germany, 2013.
4. Horvath, P.; Yampolskiy, M.; Koutsoukos, X. Efficient evaluation of wireless real-time control networks. *Sensors* **2015**, *15*, 4134–4153. [CrossRef] [PubMed]
5. Shi, Y.; Wang, J.; Fang, X.; Gu, S.; Dong, L. Modelling and control of S-MAC based wireless sensor networks control system with network-induced delay in industrial. In Proceedings of the 4th International Conference on Information Science and Control Engineering (ICISCE), Changsha, China, 21–23 July 2017; IEEE: Piscataway, NJ, USA, 2017; pp. 1539–1544.
6. Luvisotto, M.; Pang, Z.; Dzung, D. High-performance wireless networks for industrial control applications: New targets and feasibility. *Proc. IEEE* **2019**, *107*, 1074–1093. [CrossRef]
7. Park, P.; Ergen, S.C.; Fischione, C.; Lu, C.; Johansson, K.H. Wireless network design for control systems: A survey. *IEEE Commun. Surv. Tutor.* **2017**, *20*, 978–1013. [CrossRef]
8. Tipsuwan, Y.; Chow, M.Y. Control methodologies in networked control systems. *Control Eng. Pract.* **2003**, *11*, 1099–1111. [CrossRef]
9. Maass, A.I.; Nesic, D.; Postoyan, R.; Dower, P.M.; Varma, V.S. Emulation-based stabilisation of networked control systems over WirelessHART. In Proceedings of the IEEE 56th Annual Conference on Decision and Control (CDC), Melbourne, Australia, 12–15 December 2017; IEEE: Piscataway, NJ, USA, 2017; pp. 6628–6633.
10. Araújo, J.; Mazo, M.; Anta, A.; Tabuada, P.; Johansson, K.H. System architectures, protocols and algorithms for aperiodic wireless control systems. *IEEE Trans. Ind. Inform.* **2013**, *10*, 175–184. [CrossRef]
11. GODOY; Eduardo, P. *Networked Control Systems: Research Challenges and Advances for Application*; Nova Science: New York, NY, USA, 2017; ISBN 978-1-53613106-2.
12. Deng, L.; Tan, C.; Wong, W.S. On stability condition of wireless networked control systems under joint design of control policy and network scheduling policy. In Proceedings of the IEEE Conference on Decision and Control (CDC), Miami, FL, USA, 17–19 December 2018; IEEE: Piscataway, NJ, USA, 2018; pp. 2041–2047.
13. Polastre, J.; Szewczyk, R.; Culler, D. Telos: Enabling ultra-low power wireless research. In Proceedings of the Fourth International Symposium on Information Processing in Sensor Networks (IPSN 2005), Boise, ID, USA, 15 April 2005; IEEE: Piscataway, NJ, USA, 2005; pp. 364–369.
14. Ahlén, A.; Akerberg, J.; Eriksson, M.; Isaksson, A.J.; Iwaki, T.; Johansson, K.H.; Knorn, S.; Lindh, T.; Sandberg, H. Toward wireless control in industrial process automation: A case study at a paper mill. *IEEE Control Syst. Mag.* **2019**, *39*, 36–57. [CrossRef]
15. Gulati, M.K.; Kumar, K. Survey of stability based routing protocols in mobile ad-hoc networks. In Proceedings of the ICCCS International Conference on Communication, Computing & Systems, Macau, China, 19–21 November 2014; pp. 100–105.
16. Brahmbhatt, S.; Kulshrestha, A.; Singal, G. SSLSM: Signal strength based link stability estimation in MANETs. In Proceedings of the Computational Intelligence and Communication Networks (CICN), Jabalpur, India, 12–14 December 2015; IEEE: Piscataway, NJ, USA, 2015; pp. 173–177.
17. Florencio, H.; Neto, A.D.D. Method for link stability evaluation of industrial wireless sensor networks (ISA 100.11 a). *Prz. Elektrotechniczny* **2019**, *11*, 176–183. [CrossRef]
18. Zou, Y.; Tao, Y. A method of selecting path based on neighbor stability in ad hoc network. In Proceedings of the International Conference on Computer Science and Automation Engineering, Zhangjiajie, China, 14–15 January 2012; IEEE: Piscataway, NJ, USA, 2012; pp. 675–678.
19. Sun, J.; Liu, Y.A.; Hu, H.; Yuan, D. Link stability based routing in mobile ad hoc networks. In Proceedings of the 5th IEEE Conference on Industrial Electronics and Applications (ICIEA), Taichung, Taiwan, 15–17 June 2010; IEEE: Piscataway, NJ, USA, 2010; pp. 1821–1825.

20. Boukerche, A.; Coutinho, R.W.; Yu, X. LISIC: A link stability-based protocol for vehicular information-centric networks. In Proceedings of the IEEE 14th International Conference on Mobile Ad Hoc and Sensor Systems (MASS), Orlando, FL, USA, 22–25 October 2017; IEEE: Piscataway, NJ, USA, 2017; pp. 233–240.

21. De Rango, F.; Guerriero, F.; Fazio, P. Link-stability and energy aware routing protocol in distributed wireless networks. *IEEE Trans. Parallel Distrib. Syst.* **2012**, *23*, 713–726. [CrossRef]

22. Zhang, Z.; Jia, Z.; Xia, H. Link stability evaluation and stability based multicast routing protocol in mobile ad hoc networks. In Proceedings of the 11th International Conference on Trust, Security and Privacy in Computing and Communications, Liverpool, UK, 25–27 June 2012; IEEE: Piscataway, NJ, USA, 2012; pp. 1570–1577.

23. Sharma, S.; Pathak, D. Energy aware path formation with link stability in wireless adhoc network. In Proceedings of the International Conference on Advances in Computer Engineering and Applications, Ghaziabad, India, 19–20 March 2015; IEEE: Piscataway, NJ, USA, 2015; pp. 826–831.

24. Idoudi, H.; Abderrahim, O.B.; Mabrouk, K. Generic links and paths stability model for Mobile Ad Hoc Networks. In Proceedings of the International Wireless Communications and Mobile Computing Conference (IWCMC), Paphos, Cyprus, 5–9 September 2016; IEEE: Piscataway, NJ, USA, 2016; pp. 394–398.

25. Chama, N.; Sofia, R.C.; Sargento, S. Multihop mobility metrics based on link stability. In Proceedings of the 9th International Symposium on Frontiers of Information Systems and Network Applications (FINA2013), Barcelona, Spain, 25–28 March 2013; pp. 809–816.

26. Al-Qudah, Z.; Alsarayreh, M.; Jomhawy, I.; Rabinovich, M. Internet path stability: Exploring the impact of mpls deployment. In Proceedings of the Global Communications Conference (GLOBECOM), Washington, DC, USA, 4–8 December 2016; IEEE: Piscataway, NJ, USA, 2016; pp. 1–7.

27. Wang, S.; Song, Q.; Feng, J.; Wang, X. Predicting the link stability based on link connectivity changes in mobile ad hoc networks. In Proceedings of the International Conference on Wireless Communications, Networking and Information Security, Beijing, China, 25–27 June 2010; IEEE: Piscataway, NJ, USA, 2010; pp. 409–414.

28. Li, X.; Yan, J. LEPR: Link stability estimation-based preemptive routing protocol for flying Ad hoc Networks. In Proceedings of the Symposium on Computers and Communications (ISCC), Heraklion, Greece, 3–6 July 2017; IEEE: Piscataway, NJ, USA, 2017; pp. 1079–1084.

29. Priya, D.; Haripriya, K. An energy efficient link stability based routing scheme for wireless sensor networks. In Proceedings of the International Conference on Communication and Signal Processing (ICCSP), Chennai, India, 6–8 April 2017; IEEE: Piscataway, NJ, USA, 2017; pp. 1828–1832.

30. Lei, W. Path stability based source routing in mobile ad hoc networks. In Proceedings of the Cross Strait Quad-Regional Radio Science and Wireless Technology Conference (CSQRWC), Harbin, China, 26–30 July 2011; pp. 774–778.

31. Patil, R.B.; Patil, A.B. Energy, link stability and queue aware OLSR for Mobile Ad hoc Network. In Proceedings of the International Conference on Advances in Computing, Communications and Informatics (ICACCI), Kochi, India, 10–13 August 2015; IEEE: Piscataway, NJ, USA, 2015; pp. 1020–1025.

32. Walsh, G.C.; Ye, H.; Bushnell, L.G. Stability analysis of networked control systems. *IEEE Trans. Control Syst. Technol.* **2002**, *10*, 438–446. [CrossRef]

33. International Electrotechnical Commission (IEC). *Industrial Communication Networks—Wireless Communication Network and Communication Profiles-ISA 100.11 (IEC 62734)*; IEC: Geneva, Switzerland, 2012.

34. Yokogawa Electric. YFGW410 Field Wireless Management Station. Avaiable online: https://www.yokogawa.com/solutions/products-platforms/field-instruments/field-wireless/wireless-infrastructure/ (accessed on 4 July 2020).

35. Yokogawa Electric. YFGW510/YFGW520 Field Wireless Access Point. Avaiable online: https://www.yokogawa.com/solutions/products-platforms/field-instruments/field-wireless/wireless-temperature-transmitters/ (accessed on 4 July 2020).

36. Espressif. ESP32: A Feature-Rich MCU with Integrated Wi-Fi and Bluetooth Connectivity for a Wide-Range of Applications. Avaiable online: https://www.espressif.com/en/products/socs/esp32/overview (accessed on 4 July 2020).

37. Aghenta, L.O.; Iqbal, M.T. Low-cost, open source IoT-based SCADA system design using Thinger.IO and ESP32 thing. *Electronics* **2019**, *8*, 822. [CrossRef]

38. Taştan, M.; Gokozan, H. Real-time monitoring of indoor air quality with internet of things-based E-nose. *Appl. Sci.* **2019**, *9*, 3435. [CrossRef]

39. Gutierrez-Guerrero, J.M.; Holgado-Terriza, J.A. Automatic configuration of OPC UA for Industrial Internet of Things environments. *Electronics* **2019**, *8*, 600. [CrossRef]

40. Figueroa-Lorenzo, S.; Anorga, J.; Arrizabalaga, S. A role-based access control model in modbus SCADA systems: A centralized model approach. *Sensors* **2019**, *19*, 4455. [CrossRef]

41. Silva, J.P.S.D. Modelo de Sistema de Automação Aplicado à Operação de Redes de Abastecimento Hídrico. Master's Thesis, Federal University of Rio Grande do Norte, Natal, Brazil, 2019.

MDPI
St. Alban-Anlage 66
4052 Basel
Switzerland
Tel. +41 61 683 77 34
Fax +41 61 302 89 18
www.mdpi.com

Sensors Editorial Office
E-mail: sensors@mdpi.com
www.mdpi.com/journal/sensors